PHYSICS

A. J. Flint, B.Sc., M.Sc.

Published by Intercontinental Book Productions in conjunction with Seymour Press Ltd

Distributed by Seymour Press Ltd., 334 Brixton Road, London, S.W.9

1st Edition 11th Impression © Intercontinental Book Productions 1973

Printed in England by McCorquodale (Newton) Ltd.

Newton-le-Willows,

Lancashire

ISBN 0 85047 996 7

6/79/15

Contents

Key Facts 'A' Level Books

Key Facts Books form the basis for examination answers, and cover the basic areas of each subject. They are ideal for a quick grasp of the subject and for revision, but should also be used as a reference throughout the course of study.

Key Facts Books are compiled by teachers who have detailed first-hand knowledge of examining and examinations in addition to their considerable classroom experience.

Key Facts Books are suitable for students studying for G.C.E. A Level or equivalent standard examinations.

Key Facts cover the most important parts of each subject syllabus for the various Examining Boards.

Key Facts Books enable a complete subject summary to always be carried in a pocket for learning in spare moments.

Key Facts Books make study time more productive, because time available can be spent actually learning from expertly produced material.

Remember that no-one can pass examinations for you. Success depends on how much effort **you** are able to make.

Learn the facts thoroughly. Regular review, throughout your course, is far better than last minute cramming. Try to work through one section each day. **Think** about the work. Try to **understand** it. Examiners look not only for understanding of the facts, but also for their inter-relationship and explanations.

Be sure to know your syllabus. Work through past papers to practise types of questions which are most often set. Use textbooks for fuller details if this seems necessary.

**Always carry Key Facts Books with you,
and let them help you to Exam success.**

Mechanics

DIMENSIONS

The basic dimensions in mechanics are mass (M), length (L) and time (T).
Velocity, = rate of change of displacement, has dimensions LT^{-1}.
Acceleration, = rate of change of velocity, has dimensions LT^{-2}. Force, = mass × acceleration, has dimensions MLT^{-2}. The two sides of an equation must be both numerically and dimensionally identical. Dimensions may be used to check the form of an equation or to derive the form of a relationship.

Example. The time of swing of a simple pendulum may depend on the mass of the bob (m), the length of string (l) and the acceleration due to gravity (g).
Time $\propto$ (mass)^x × (length)^y × (acceleration due to gravity)^z
Dimensionally, $T^{+1} = [M]^x.[L]^y.[LT^{-2}]^z$
Equating powers of M, $0 = x$
Equating powers of L, $0 = y + z$
Equating powers of T, $1 = -2z$
Therefore $z = -\frac{1}{2}$ and $y = +\frac{1}{2}$. The time is independent of m.
Time $\propto \sqrt{l/g}$
Non-dimensional constants, 2π in this case, cannot be obtained in this way.

VECTORS AND SCALARS

Scalar quantities have magnitude only. A vector quantity has both magnitude and direction associated with it.
The product of a vector and a scalar is always a vector, e.g. mass × velocity gives momentum which is a vector.
The product of two vectors may be a scalar or a vector, e.g. kinetic energy ($\frac{1}{2}mv^2$) is a scalar where v is a vector.
Vectors are added by the parallelogram or triangle law of vector addition. To subtract two vectors, $\vec{V}_1$ and $\vec{V}_2$, note that $\vec{V}_1 - \vec{V}_2 = \vec{V}_1 + (-\vec{V}_2)$. In other words, the procedure is to reverse the vector which is being subtracted and then add it using the parallelogram or triangle construction. An example will be found in the section on motion in a circle.

MOMENTS AND COUPLES

The **moment of a force about an axis** is the **product of the force and the perpendicular distance between the force and the axis.** This **moment** is a **vector quantity,** and the vector is taken to act along the axis in a direction given by a right hand corkscrew rule. If two turning moments are applied to the same body, the resultant axis of rotation and the magnitude of the resultant moment are determined by vector addition of the two axial vector moments.

When **two equal and opposite forces,** not acting in the same line, are applied to a body, it tends to rotate. This arrangement of forces is called a **couple.** A couple cannot be replaced by a single force, but can be replaced by any other couple having the same moment and acting in the same plane. The moment of a couple is the product of one force and the perpendicular distance between the forces, and is called the **torque.**

CENTRES OF GRAVITY AND OF MASS. EQUILIBRIUM

The **centre of gravity** of a body is the **point where the resultant force of attraction (weight) of the body acts.** The **centre of mass** is the **point where the total mass of the body acts or appears to act.** These points coincide except in the case of a very large body over which g is not uniform.

A single force which does not act through the centre of mass of a ball, causes it to spin as well as move forward. In such a case two equal and opposite forces may be thought of as acting through the centre of mass, parallel with the original force. This corresponds to a forward force acting through the centre of mass, together with a couple which causes rotation. If a body is in **equilibrium,** then the **resultant force** and the **resultant couple** must **both be zero.**

Two forces in equilibrium must be equal and opposite.

For three force equilibrium, the **resultant of any two of the forces must be equal and opposite to the third force,** and all three forces must act in the same plane. If two of the forces are parallel, their resultant will be parallel to them, hence all three forces must be parallel. If the **forces are not all parallel,** then **they must act through a single point.**

6

SENSITIVITY OF A BEAM BALANCE

A beam balance effectively compares masses, because the value of g is constant over the region occupied by the balance. The instrument is **sensitive** if a small change in the mass in one pan causes a large change in deflection

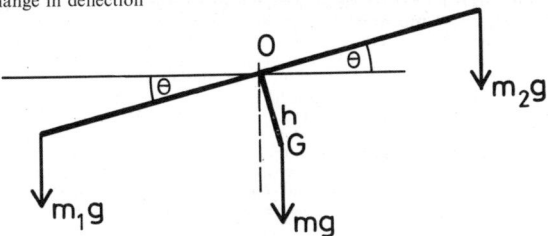

Figure 1

Suppose the mass of the beam is m, and its centre of mass is at G, a distance h below a line through the knife edges which support the pans. The length of each arm is l. A mass $m_1(> m_2)$ causes a deflection θ. Taking moment about O:

$$m_1 gl \cos \theta = mgh \sin \theta + m_2 gl \cos \theta$$

$$\therefore (m_1 - m_2)l \cos \theta = mh \sin \theta$$

$$\therefore \tan \theta = \frac{(m_1 - m_2)l}{mh}$$

A sensitive balance has a long beam and small values of m and h. In practice problems of rigidity also affect the design and the values of these quantities.

BUOYANCY CORRECTIONS

A kilogramme of lead and a kilogramme of feathers must, by definition, balance each other in a vacuum. But in air each will experience an upthrust equal to the weight of fluid displaced **(Archimedes' principle).**

For a body of mass m and density ρ, the volume of air displaced is m/ρ, and the weight of this air is mdg/ρ, where d is the density of air. The net downthrust is $mg - mdg/\rho$.

So for balance, $m\left(1 - \dfrac{d}{\rho}\right) = m'\left(1 - \dfrac{d}{\rho'}\right)$.

DYNAMICS

For uniformly accelerated motion with the usual symbols: $s = \frac{1}{2}(u + v) . t$, $v = u + at$, $s = ut + \frac{1}{2}at^2$, $v^2 = u^2 + 2as$. **Acceleration** is equal to the **slope of a velocity-time graph**, and **distance travelled** is equal to the **area between the graph and the time axis**.

MOTION UNDER GRAVITY IN TWO DIMENSIONS

In the absence of friction and air resistance, the only force acting on a body in free flight is its weight. This acts vertically downwards, and since there is no horizontal force, there is no horizontal acceleration. The vertical component of motion is independent of the horizontal component.

Example. A stone is thrown horizontally from the top of a vertical cliff which is 20 m high with a velocity of 12 m s^{-1}. Find how far it lands from the base of the cliff and also its velocity on impact with the ground. Take $g = 10$ m s^{-2}.

Answer. For the vertical component of motion, $u = 0$.
$s = ut + \frac{1}{2}at^2$ becomes $20 = 0 + \frac{1}{2} . 10 . t^2$. Hence $t = 2$ s. With a uniform horizontal velocity of 12 m s^{-1}, the stone will travel 24 m in 2 s, and this is the answer to the first part. The vertical component of the stone's final velocity is $v = at = 20$ m s^{-1}. To this must be added, by vector addition, the horizontal velocity of 12 m s^{-1}, so the resultant is $\sqrt{20^2 + 12^2} = 23 \cdot 3$ m s^{-1} at an angle of $\tan^{-1} 20/12$ below the horizontal.

In general, for a projectile given an initial velocity of V at an angle θ with the ground:
For the upward flight: $u = V \sin \theta$, $v = 0$ and $a = -g$.
$v = u + at$ gives $0 = V \sin \theta - gt$, so $t = V \sin \theta/g$.

$s = \frac{1}{2}(u + v) . t = \dfrac{V^2 \sin^2 \theta}{2g}$ gives the height of the trajectory.

The time of flight, $T = 2t$.
The constant horizontal velocity is $V \cos \theta$.

Hence the range is $V \cos \theta . T = \dfrac{2V^2 \sin \theta . \cos \theta}{g} = \dfrac{V^2 \sin 2\theta}{g}$

The maximum range occurs when $\sin 2\theta = 1$, i.e. $\theta = 45°$, and the value for the maximum range is then V^2/g.

NEWTON'S LAWS OF MOTION

1. **A body will continue in a state of rest or uniform motion in a straight line unless acted upon by a resultant force.**
2. **The rate of change of momentum of a body is proportional to the impressed force and takes place in the direction of the force.**

Force $\propto$ rate of change of momentum $= \dfrac{mv - mu}{t} = ma$ where m is the inertial mass of the body.

In SI the unit of force, the newton, is defined as that force required to give a mass of 1 kg an acceleration of 1 m s^{-2}.
An advantage of this definition is that force = rate of change of momentum, or $F = ma$.
The weight of a body is a measure of the force of gravity on it.
Weight = mass × acceleration due to gravity is a special case of the equation $F = ma$. The value of g varies from place to place, causing the weight of a standard mass to vary consequentially. On earth, the weight of a standard kilogramme is about 9·8 N, but on the moon its weight would be about 1·6 N. The acceleration due to gravity, g, is normally measured in m s^{-2}, but since g = weight/mass, it may also be thought of as a measure of the earth's gravitational field strength in N kg^{-1}.

Since, in SI, $F = \dfrac{mv - mu}{t}$, then $Ft = m(v - u)$. Ft is called the impulse and it is equal to the change in momentum. If two bodies, A and B, collide and are in contact for t s, then there is a force between them for t s. Newton's third law says:

3. **To every action there is an equal and opposite reaction.**

The average force exerted by A on B (action) is equal to the average force exerted by B on A (reaction), but the two forces act in different directions.
The change in momentum of A is $Ft = \Delta(mv)_A$.
The change in momentum of B is $-Ft = -\Delta(mv)_B$.
Hence the total change in momentum $= \Delta(mv)_A + -\Delta(mv)_B = 0$, a result which may be stated as the **law of conservation of momentum**, i.e. **in a system to which no external force is applied, the total momentum of the system must remain constant.** Conservation of linear momentum applies to all collision and explosive processes.

Example. If a radioactive nucleus of atomic mass number 238 emits an alpha particle (of mass number 4) with velocity v, then the recoil velocity of the nucleus, V, is given by: -
$(4v) + (-234V) = 0$. Hence $V = 4v/234$.
Where an **oblique collision** occurs, the direction of the incident particle may be taken as a reference line and all momenta are resolved into components parallel and perpendicular to this direction. Conservation must then apply in both directions.

Example. A body of mass 2 kg travelling due east with a velocity of 4 m s^{-1} collides with a body of mass 3 kg travelling due north with a velocity of 6 m s^{-1}. If the bodies coalesce on impact, find their common velocity.

Answer. Momentum in W-E direction: $2 \times 4 = 5 \times V_e$
Momentum in S-N direction: $3 \times 6 = 5 \times V_n$
The combined body thus has a velocity of 1·6 m s^{-1} eastwards, together with a velocity of 3·6 m s^{-1} northwards. These two components may be combined (by vector addition) to give the magnitude and direction of the resultant.

WORK AND ENERGY

The work done by a force is equal to the product of the force and the distance moved in the direction of the force. If the force acts at an angle θ to the direction of the resulting motion, as in the case of pushing a lawnmower, then the work done is $F \cdot \cos\theta \times$ distance moved. If the equation $F = ma$ is combined with the equation of motion $v^2 = u^2 + 2as$ by the elimination of a, then the result may be written as $Fs = \frac{1}{2}mv^2 - \frac{1}{2}mu^2$. This equation indicates that the work done by a force is equal to the change in kinetic energy produced. The work done in lifting a weight ($W = mg$) to a height h above the earth is mgh, and is equal to the gain in potential energy. Energy and work are both **scalar** quantities.
Energy cannot be created or destroyed. It may be converted partially or wholly from one form to another. Collisions are of two kinds, elastic, in which no kinetic energy is converted to any other form, and inelastic collisions, in which kinetic energy is not conserved, much of the "lost" K.E. being changed into low-grade heat energy. Collisions between gas molecules appear to be elastic, otherwise cooling would be observed in an isolated gas.

CIRCULAR MEASURE

One **radian** is the angle subtended at the centre of a circle by an arc of the circle equal in length to the radius. $360°$ is equivalent to 2π radians. An arc of length s subtends an angle θ at the centre of a circle of radius r such that $\theta = s/r$ radians, so $s = r\theta$.

If a body moves a distance s along a circular arc in time t, then its linear speed $v = s/t = r\theta/t = \omega r$ where ω is the angular speed in radian$.\text{s}^{-1}$ and $v = r\omega$.

WORK DONE BY A COUPLE

If two forces of magnitude F separated by a distance $2r$ together constitute a couple which rotates through an angle θ, then the work done by the couple is $F.2r.\theta$, i.e. couple $\times$ angle of rotation in radians. Power is equal to the rate of working $= 2rF\theta/t = 2rF\omega$. Note that the units of rotational work and power are the same as those for translational motion, joules and watts respectively.

CIRCULAR MOTION

A stone attached to a string and whirled round at constant speed requires a force to maintain its path. This force is called the centripetal force and acts towards the centre of the circle. The presence of this force implies that there will be an acceleration also acting towards the centre of the circle.

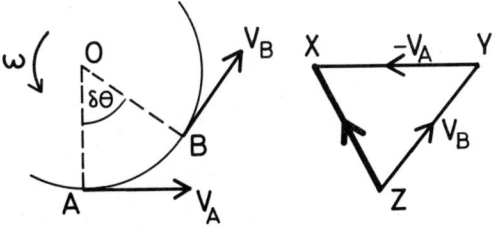

Figure 2

11

Consider a particle of mass m moving with steady speed in a circle of radius r. At A its velocity is $\vec{v}_A$ and at B its velocity is $\vec{v}_B$. The velocity change is $\vec{v}_B - \vec{v}_A = \vec{v}_B + (-\vec{v}_A) = \vec{ZX}$. If the particle moves through an angle $\delta\theta$ in time δt, then the angular velocity, $\omega = \delta\theta/\delta t$. If δt is small, then $\delta\theta$, which is equal to the angle XYZ, is also small. $\vec{ZX}$ then points to the centre of the circle, and is of length $v.\delta\theta$. The acceleration of the particle, a, is given by

$$a = \frac{\text{change in velocity}}{\text{time}} = \frac{ZX}{\delta t} = \frac{v.\delta\theta}{\delta t}$$

In the limit, as δt tends to zero, $\delta\theta/\delta t = d\theta/dt = \omega$.
Hence the acceleration $= v\omega$, or v^2/r or $r\omega^2$.
The centripetal force $=$ mass $\times$ acceleration $= mv^2/r = mr\omega^2$.
If a stone is whirled in a **vertical circle** with constant angular velocity, the centripetal force at all points must be equal to $mr\omega^2$. The tension in the string may also depend on the weight of the stone, mg, which always acts vertically downwards.
At 9 or 3 o'clock positions, tension $= mr\omega^2$ since this tension acts at right angles to the weight.
At 12 o'clock, tension $= mr\omega^2 - mg$.
At 6 o'clock, tension $= mr\omega^2 + mg$.
In the latter cases note that the tension is the force exerted by the stone on the string, while the centripetal force is exerted by the string on the stone. The string is most likely to break when the stone is at the 6 o'clock position, i.e. at the bottom of its path.
The **conical pendulum** consists of a bob on the end of a light inextensible string. In equilibrium the string hangs vertically, but if the bob is caused to move in a horizontal circle, with uniform angular velocity, the string makes an angle α with the vertical. If the tension in the string is T, then:
Resolving vertically: $T \cos\alpha = mg$.
Resolving horizontally: $T \sin\alpha = mr\omega^2 = ml \sin\alpha.\omega^2$ where l is the length of the string.

$$\therefore \frac{T \sin\alpha}{T \cos\alpha} = \frac{ml \sin\alpha.\omega^2}{mg} \quad \text{or} \quad \omega^2 = \frac{g \tan\alpha}{l \sin\alpha} = \frac{g}{l \cos\alpha}.$$

The **periodic time,** $T, = 2\pi/\omega = 2\pi\sqrt{\dfrac{l \cos\alpha}{g}}.$

MOMENT OF INERTIA

If a heavy, rigid body rotates about an axis with uniform angular velocity, then all the individual particles making up the body have a common angular velocity, ω, but they all have different values of v and r. For each particle, $v = r\omega$. The kinetic energy of a typical particle = $\frac{1}{2}mv^2 = \frac{1}{2}mr^2\omega^2$. The total kinetic energy of the body is $\frac{1}{2}\omega^2$(sum of all mr^2) = $\frac{1}{2}\omega^2 \Sigma mr^2$.

$\Sigma mr^2 = I$, **the moment of inertia** of the body about the axis. A large moment of inertia means that it is difficult to increase the angular velocity of the body. I depends on the mass of the body but also upon the distribution of the mass. The same body will have different values of I depending on which axis it rotates about.

In general $I = Mk^2$ where k is a constant called the **radius of gyration,** and M is the total mass of the body. I is always proportional to M and to the square of a distance.

I for a uniform rod about an axis through the centre = $Ml^2/12$.

I for a uniform rod about an axis through one end = $Ml^2/3$.

I for a circular disc about an axis through its centre and perpendicular to its plane = $Ma^2/2$, where a is the radius.

I for a sphere about an axis through its centre = $2Ma^2/5$.

For a wheel rolling along the ground, total kinetic energy = $\frac{1}{2}Mv^2 + \frac{1}{2}I\omega^2$. The simplicity of this equation illustrates the scalar nature of energy.

I is the constant of proportionality between the moment of a force or couple (torque) acting on a body and the angular acceleration produced. If T is the torque and α the angular acceleration, then the **equations of uniform rotational acceleration are analogous to those of uniform linear acceleration.**

Linear Motion	Rotational Motion
$v = u + at$	$\omega_2 = \omega_1 + \alpha t$
$F = ma$	$T = I\alpha$
$Ft = mv - mu$	$Tt = I\omega_2 - I\omega_1$

The product $I\omega$ is known as the **angular momentum.**

$Fs = \frac{1}{2}mv^2 - \frac{1}{2}mu^2$ $T\theta = \frac{1}{2}I\omega_2^2 - \frac{1}{2}I\omega_1^2$

Note that α is measured in radians s^{-2}. The principles of **conservation of angular momentum** and angular kinetic energy are analogous with those of linear dynamics.

SIMPLE HARMONIC MOTION

A particle P moves round a circle of radius a with uniform angular velocity ω. If the time taken to move from X to P is t s, then

$$\omega = \frac{\text{angle } POX \text{ in radians}}{t}$$

Hence angle $POX = \omega t$. N is the foot of the

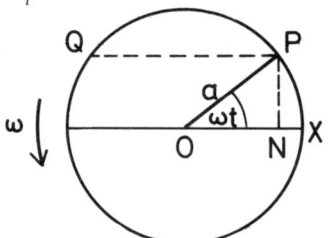

Figure 3

perpendicular drawn from P to OX. The velocity of the particle is $a\omega$ tangential to the circle. This velocity may be resolved into two components, $a\omega \cos \omega t$ along NP and $a\omega \sin \omega t$ along PQ. But since N is always vertically below P, the velocity of N along the diameter of the circle is the same as this latter resolved component of the velocity of P. Since N is moving towards O, its velocity in the positive direction of X is $-a\omega \sin \omega t$.

Hence acceleration of $N = dv/dt = -a\omega^2 \cos \omega t$.

But if the displacement $ON = x$, then $x = a \cos \omega t$, so acceleration of $N = -\omega^2 x$.

The point N executes S.H.M. about O.

Simple harmonic motion may be defined by two statements:

1. **The acceleration of a particle moving with S.H.M. is always directed towards a fixed point.**
2. **The acceleration is directly proportional to the displacement of the particle from that fixed point.**

The latter statement arises since ω, and hence ω^2, is constant. One complete oscillation of N is equivalent to one complete circle by P, i.e. 2π radians. If T is the time taken for one complete oscillation, then $T = 2\pi/\omega$. T is known as the **periodic time**, or **period**, of the motion.

a is called the amplitude of the oscillation and is equal to the maximum displacement of the particle.

The period $T = 1/f$, where f is the **frequency in hertz.**

At any time t, the velocity v of the particle is $v = -a\omega \sin \omega t$. Hence $v^2 = a^2\omega^2 \sin^2 \omega t = a^2\omega^2(1 - \cos^2 \omega t) = \omega^2(a^2 - x^2)$

$$\therefore v = \pm\omega\sqrt{a^2 - x^2}$$

The maximum velocity occurs when $x = 0$. $v_{max} = \pm a\omega$.

The maximum acceleration (when $x = \pm a$) is $\pm a\omega^2$.

The Simple Pendulum. If the bob is displaced from the vertical through an angle θ, the weight of the bob may be resolved into two components, $mg \cos \theta$ parallel to the string, and $mg \sin \theta$ perpendicular to the string. This latter force, together with an equal and opposite reaction at the support, constitute an accelerating couple. The restoring torque is $-mgL \sin \theta$, where L is the length of the string.

Angular acceleration $\alpha = \dfrac{T}{I} = \dfrac{-mgL \sin \theta}{mL^2} = \dfrac{-g \sin \theta}{L}$.

For small displacement θ is small, and $\sin \theta \to \theta$.

Angular acceleration $= -g\theta/L = -\omega^2\theta$. Hence $\omega^2 = g/L$.

The period $T = \dfrac{2\pi}{\omega} = 2\pi\sqrt{\dfrac{L}{g}}$.

The Helical Spring. The ratio load/extension $= \lambda$, a constant. When a mass is hung from a vertical spring, it stretches by an amount e, given by $mg = \lambda e$. If the load is pulled down a further distance x, the new tension in the spring is $\lambda(e + x)$.

When the load is released the net upward force is λx.

Using $F = ma$, $a = \dfrac{F}{m} = -\dfrac{\lambda}{m} \cdot x = -\omega^2 x$. $\qquad \therefore T = \dfrac{2\pi}{\omega} = 2\sqrt{\dfrac{m}{\lambda}}$.

The energy of the spring is potential. P.E. for an extension $y = \int F \cdot dy = \int \lambda y \cdot dy = \frac{1}{2}\lambda y^2$.

The energy of the mass is kinetic ($\frac{1}{2}mv^2$). The displacement, $y = a \cos \omega t$, and the corresponding velocity $= -a\omega \sin \omega t$.

Total energy of system $=$ P.E. of spring $+$ K.E. of mass.

$= \frac{1}{2}\lambda y^2 + \frac{1}{2}mv^2 = \frac{1}{2}m\omega^2 y^2 + \frac{1}{2}mv^2$

$= \frac{1}{2}m\omega^2 a^2 \cos^2 \omega t + \frac{1}{2}m\omega^2 a^2 \sin^2 \omega t = \frac{1}{2}m\omega^2 a^2$.

This expression is independent of y, and so is constant throughout the oscillation. This is a general result for S.H.M.

15

GRAVITATION
Kepler's Laws:

1. The planets move round the sun in elliptical orbits with the sun at one focus.
2. Each planet revolves in such a way that an imaginary line joining it to the sun sweeps out equal areas in equal times.
3. The squares of the times of revolution of the planets are proportional to the cubes of their mean distances from the sun.

Newton's Law of Gravitation:

The gravitational force of attraction between two masses is proportional to the product of their masses and inversely to the square of the distance between their centres of mass.

$F = G\dfrac{m_1 m_2}{r^2}$ G is the universal gravitational constant. Its value is $6 \cdot 67 \times 10^{-11}$ N m^2 kg^{-2}.

At the surface of the earth, the force on a mass $m = mg$. So $mg = G\dfrac{Mm}{r^2}$ where M is the mass of the earth and r is the radius. $\therefore g = GM/r^2$. Since G and M are constant, g depends on $1/r^2$. The earth has a greater radius at the equator than at the poles, and this is one reason for variations in g over the surface of the earth.

The earth rotates once every 24 hours, and a particle at rest on the equator is moving in a circle of radius about 6 400 km. Its speed is $2\pi(6 \cdot 4 \times 10^6)/24 \times 60 \times 60$ m s^{-1}, and its centripetal acceleration (v^2/r) is $0 \cdot 3$ m s^{-2}. If g for a freely falling body were measured in a laboratory on the equator, everything in the laboratory would have the calculated value of this centripetal acceleration, and only the "left over" value of the earth's gravitational field strength would be measured. At the poles, however, there is no centripetal acceleration and the whole pull of gravity would cause the acceleration of the falling body. This is the other main reason for variations in g over the surface of the earth. When the earth was still a mobile material, the higher pull on materials in the polar regions compared with that at the equator caused the earth to take the shape of an oblate spheroid. Knowing G, r, and g, the equation $g = GM/r^2$ may be used to calculate a value for the mass of the earth, M. Taking it to be a sphere, the average density may then be calculated, $(V = \frac{4}{3}\pi r^3)$.

SATELLITES

For any earth satellite, including the moon, the centripetal force needed to maintain orbit is provided by the earth's gravitational pull.

$mR\omega^2 = G\dfrac{Mm}{R^2}$ where R is the radius of the orbit. R is equal to $r + h$, the radius of the earth + the height of the satellite.

$$\therefore mR\left(\frac{2\pi}{T}\right)^2 = G\frac{Mm}{R^2} \quad \text{or} \quad \frac{4\pi^2 R}{T^2} = G\frac{M}{R^2}$$

$$\therefore T^2 = \frac{4\pi^2 R^3}{GM}, \text{ i.e. } T^2 \propto R^3 \text{ (cf Kepler's third law.)}$$

Note that, for a given orbit, the speed of the satellite is independent of its mass. As the radius of the orbit decreases the speed of the satellite increases. This does not contradict the law of conservation of energy since the increased kinetic energy of the satellite occurs at the expense of its gravitational potential energy.

If a satellite is in orbit above the equator and travelling in the same direction as the earth rotates, then if $T = 24$ hours, the satellite will remain directly above the same point on the earth's surface. It is said to be in **synchronous orbit**.

$$R = \sqrt[3]{\frac{T^2 GM}{4\pi^2}} = \sqrt[3]{\frac{T^2 gr^2}{4\pi^2}} \quad \text{since } GM/r^2 = g$$

$$R = \sqrt[3]{\frac{(24 \times 60 \times 60)^2 \times 9\cdot8 \times (6\cdot4 \times 10^6)^2}{4\pi^2}} = 4\cdot2 \times 10^7 \text{ m}$$

Variation of g with height. The acceleration due to gravity decreases with height above the earth's surface.

$mg = G\dfrac{Mm}{r^2}$ and $mg' = G\dfrac{Mm}{R^2}$ where $R = r + h$, the height of the satellite, and g' is the acceleration due to gravity at this height. Hence $g' = gr^2/R^2 = gr^2/(r + h)^2 \simeq g(1 - 2h/r)$ if $h \ll r$.

Weightlessness. An astronaut in a space vehicle in orbit around the earth becomes weightless since there is no reaction at the floor of the vehicle acting on him. Both astronaut and vehicle are in free fall around the earth.

Another type of weightlessness occurs when the gravitational pull of the earth is exactly balanced by that of the moon, giving rise to a gravitational null point in space.

ELASTICITY

A force applied to a material acts on an array of molecules. Any deformation of the material which takes place depends on the spread of the force over the array, i.e. on the applied **stress,** which is defined as force per unit area. The resulting deformation is measured in terms of the change in dimension/unit dimension and is called the **strain.**
For a rod or wire, stretched longitudinally:
Normal stress, σ = force/unit area (F/A).
Linear strain, ε = extension/original length (e/l).
The **modulus of elasticity, Young modulus = stress/strain**
$E = F/A \div e/l = Fl/Ae$. Dimensions are those of stress, since strain is dimensionless. The units of E are N m^{-2}.

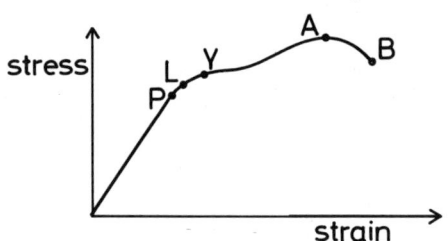

P = limit of proportionality
L = elastic limit
Y = yield point
A = breaking stress
B = sample breaks

Figure 4

If the deformation increases with time for a fixed stress the material flows and when the stress is removed the material will not revert to its original shape. It is said to be **perfectly elastic** if it does regain its shape and the maximum stress which can be applied for this to be true is called the **elastic limit.** E is constant (straight line graph) up to P, the **limit of proportionality.** P and L are close, but do not necessarily coincide. All the work done in deforming a perfectly elastic material is recoverable, being stored as potential, or strain, energy. If the elastic limit, but not the **yield point** Y is exceeded, then on removing the stress the material will only partly regain its shape and will have a **permanent set.** Once the yield point is passed, plastic flow takes place and the extension is permanent. When the breaking stress is reached, at A, the material flows rapidly thus reducing the stress until, at B, the material breaks.

Young modulus may be determined as the slope of the straight part of the stress/strain graph. $E = \sigma/\varepsilon$.

The **work done** in extending a wire = force × distance = average force × extension = $\frac{1}{2}Fe$.

This is **equal to the area under a force/extension graph.**
Up to the elastic limit, work done = strain energy stored.

Thus energy stored = $\frac{1}{2}Fe = \dfrac{1}{2}\dfrac{F}{A} \times \dfrac{e}{l} \times Al = \frac{1}{2}\sigma\varepsilon \times Al$.

But Al is the volume of the wire, so

Energy stored per unit volume = $\frac{1}{2}$ stress × strain.
This is **equal to the area under a stress/strain graph.**

Force due to expansion or contraction. If a bar of length l and linear expansivity α is heated through a temperature rise of θ K, the extension $e = \alpha l\theta$. If a force F is then applied to prevent the bar contracting,

$$F = \frac{EAe}{l} = \frac{EA\alpha l\theta}{l} = EA\alpha\theta$$

This expression may also be thought of as the force due to the expansion of the bar.

Interpretation of Stress/Strain graph. The restoring force which opposes the stretching of a sample is due to molecular attractions within the wire. Hence the elastic force has the same origin as the forces which caused the organisation of the crystals when the material solidified. If the layers of a crystal are separated slightly by a distorting force, then up to the elastic limit these molecular forces pull the layers back to their original positions once the force is removed. However, above the elastic limit the stretching force is large enough to cause **dislocations,** which are crystal imperfections, to move in the direction of the force. This causes a slip of one section of the crystal lattice, the force required to do this being less than that required to pull the entire layer forward through the same distance. The movement of these dislocations gives the wire its permanent set.

Bulk Modulus of Fluids. The **bulk stress** on a fluid is **force/area.** The force or pressure exerted on a fluid causes a change in volume, and **bulk strain = change in volume/original volume.** The **bulk modulus,** K = bulk stress/bulk strain.

SURFACE TENSION

A liquid surface is in a state of tension. The attractive force between neighbouring liquid molecules is greater than that between the liquid molecules and those of the vapour and air above the liquid. If a molecule in the surface were displaced upwards, it would be attracted from below by other liquid molecules, and energy would be required to separate the surface molecules from those below them. If a molecule in the bulk of the liquid moves to the surface, some of the bonds between it and its neighbours must be broken. **Molecules in the surface of a liquid thus have more potential energy than those in the main bulk of the liquid.**

Assuming the liquid molecules to be evenly spaced and instantaneously at rest, the average force at right angles to the line drawn on the surface will be proportional to the length of that line. The **surface tension** is defined as the **force/unit length acting perpendicular to a line drawn in the surface.** γ is measured in N m^{-1}. Thus force, $F = \gamma L$. This force in the surface of a liquid is responsible for the contraction to the smallest surface area for a particular volume of a liquid, i.e. the tendency of liquids to form spherical drops.

Angles of Contact

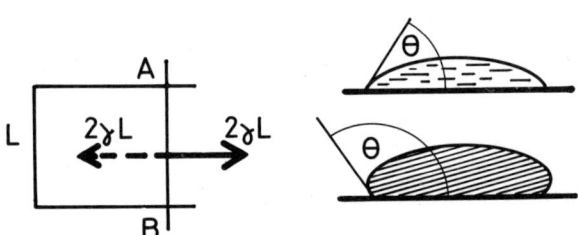

Figure 5

Work done increasing the area of a liquid surface. A film of liquid stretched across a rectangular frame is bounded by a movable wire AB of length L. The force on the wire is $2\gamma L$, the factor 2 arising because the film has two surfaces. If the wire is moved a small distance dx against the surface tension forces, the work done is $2\gamma L \cdot dx$. The increase in surface must occur under isothermal conditions, otherwise γ will change.

But $2L . dx$ is the increase in area of the film, so:

Work done $= \gamma \times$ increase in area of film.

Thus γ may also be defined as the energy required to increase the area of a liquid by unit area under isothermal conditions.

Note that $J \, m^{-2} \equiv N \, m^{-1}$.

Liquid—Solid Boundaries. When equilibrium is reached the shape of a solid-liquid boundary is such that the resultant of all the forces acting on a liquid particle at the boundary (including gravity) is zero. The angle of contact, θ, is measured between the tangent to the liquid surface at the boundary and the solid. It is measured through the liquid. If the angle of contact is zero, or nearly zero, the liquid wets the solid. If $\theta < 90°$ the meniscus is concave and if $\theta > 90°$ the meniscus is convex.

Excess Pressure in Bubbles. (a) A soap bubble has two surfaces, inside and outside. If p is the excess pressure inside a bubble of radius r, then $\pi r^2 p = 2 \times \gamma . 2\pi r \therefore p = 4\gamma/r$

(b) An air bubble inside a liquid, or a liquid drop, has one surface. $\therefore \pi r^2 p = \gamma . 2\pi r \therefore p = 2\gamma/r$

Capillary Rise. The meniscus is a segment of a sphere, and if the angle of contact is θ and the radius of curvature of the surface is R, then $R = r/\cos \theta$ where r is the internal radius of the capillary tube. The pressure just above the meniscus is 1 atmosphere, while that just below is 1 atmosphere − the pressure due to the excess column of liquid, $h\rho g$. Thus the excess pressure $= h\rho g = 2\gamma/R$

$$\therefore h\rho g = \frac{2\gamma \cos \theta}{r} \qquad \therefore \gamma = \frac{h\rho g r}{2 \cos \theta}$$

Jaeger's Method of determining γ. As water is run into a flask the displaced air moves past a manometer to escape through a capillary tube dipping to a depth h_2 vertically into the liquid. The maximum air pressure, when the bubble radius is equal to the capillary tube radius, is determined from the manometer reading. The excess pressure $(h_1 \rho_1 - h_2 \rho_2)g = 2\gamma/r$, where h_1 and ρ_1 are the height and density of the manometric liquid and h_2 and ρ_2 the height and density of the liquid under investigation. r is the radius of the capillary tube. Hence γ may be found. The method is good because the liquid in contact with the air bubble is always clean. The variation of γ with temperature may readily be determined.

Heat

THERMOMETRY

Temperature is defined by the **ideal gas equation,** $PV = RT$. P, V, and T are the pressure, volume and temperature of the gas respectively. For 1 mole of gas, R is constant **(the molar gas constant)** for all gases and has the value $8.314 \ \text{J mol}^{-1} \ \text{K}^{-1}$. In a constant volume gas thermometer, a fixed mass of gas is held at constant volume, and changes in the temperature of the gas are measured by determining the corresponding changes in pressure. Thus $P = (R/V)T$. The ideal gas equation implies an **absolute zero of temperature,** since T will equal zero when P equals zero. Absolute zero is taken as one of the fixed points on the **thermodynamic scale of temperature,** the other fixed point being the triple point of water, which, by international agreement is defined as $273.16 \ \text{K}$. This is very close to the temperature at which ice melts under a pressure of 1 atmosphere, i.e. $273 \ \text{K}$. At 'A' level these two temperatures are taken to be the same, i.e. $0°\text{C} = 273 \ \text{K}$. Thus any unknown temperature may be calculated from the relationship:

$$\frac{P_T}{P_{273}} = \frac{T}{273}$$

Where P_T is the pressure at temperature T. The unit of temperature is the kelvin.

Other scales of temperature may depend on some other property of matter which varies with temperature. Such properties include electrical resistance, thermo-electric e.m.f. and, of course, the volume of mercury in a mercury-in-glass thermometer. These scales generally have the ice point (melting point of ice under a pressure of 1 standard atmosphere), and the steam point (temperature of steam above boiling water at a pressure of 1 standard atmosphere), as their two fixed points. On the Celsius scale, the fundamental interval is divided into 100 equal degrees. The size of the Celsius degree is equal to that used in the Kelvin scale. Note that the kelvin is a fundamental SI unit equal to the fraction $1/273.16$ of the thermodynamic temperature of the triple point of water.

If X_0 is the value of the property (which varies with temperature) at the ice point, and X_{100} is its value at the steam point, then

$$\frac{\theta}{100} = \frac{X - X_0}{X_{100} - X_0}$$

where X is the value of the property at θ °C.

Note that different types of thermometer, such as mercury-in-glass and platinum resistance, will only agree at the fixed points because the assumption that these properties vary regularly with temperature is not entirely true.

CALORIMETRY

The **heat capacity** of a body, C, is the quantity of heat required to raise the temperature of the body by 1 degree. The unit of heat capacity is $J\,K^{-1}$.

The **specific heat capacity** of a substance, c, is the quantity of heat required to raise the temperature of unit mass of the substance by 1 degree without change of state. It is expressed in $J\,kg^{-1}\,K^{-1}$.

The **specific latent heat of fusion** is the heat required to convert unit mass of a solid to a liquid without a change in temperature. The unit is $J\,kg^{-1}$.

The **specific latent heat of evaporation** is the heat required to convert unit mass of a liquid to a vapour without a change in temperature. The unit is $J\,kg^{-1}$.

DETERMINATION OF SPECIFIC HEAT CAPACITIES

Specific heat capacities may be determined by:
1. Electrical methods
2. The method of mixtures
3. Methods based on Newton's law of cooling.

If heat energy q J is supplied to a body of mass m and specific heat capacity c, then the rise in temperature θ is given by $q = mc\theta$. The heat may be supplied by an electric coil, and the electrical energy either measured directly with a joulemeter, or as the product of the current in amperes, the potential difference in volts, and the time in seconds.

In general $q = mc\theta$ + heat lost to surroundings.

It is essential to compensate or correct for heat loss, and the following experiments illustrate various ways of doing this.

CONTINUOUS FLOW CALORIMETER

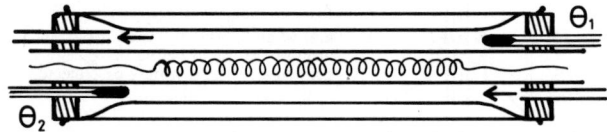

Figure 6

The liquid flows from a constant head device over an electrically heated coil. The temperatures of the liquid as it enters and leaves the coil, θ_1, θ_2, are measured when they have become steady. The mass m of liquid leaving the apparatus in t seconds is measured. If c is the mean specific heat capacity of the liquid, then $q = mc(\theta_2 - \theta_1)$ joules. This energy is provided by the electricity supply to the coil. If I is the current through the coil and V the potential difference between its ends, then

$$IVt = mc(\theta_2 - \theta_1) + \text{heat losses}$$

The heat losses depend upon the values of θ_1 and θ_2, and on the time t. To eliminate these heat losses the experiment is repeated with the rate of flow increased, and the current increased in such a way that θ_1 and θ_2 are the same as before. The new mass collected in the same time is m'. Then $I'V't = m'c(\theta_2 - \theta_1) + \text{heat losses}$.
The heat losses may be eliminated by subtracting the two equations. The specific heat capacity of the liquid may then be calculated.
Note that when the temperatures are steady, no heat is taken in warming the apparatus, since every part of it is at a constant temperature. The heat capacity of the apparatus is not required, a principal advantage of the method. The vacuum surrounding the liquid reduces heat losses.

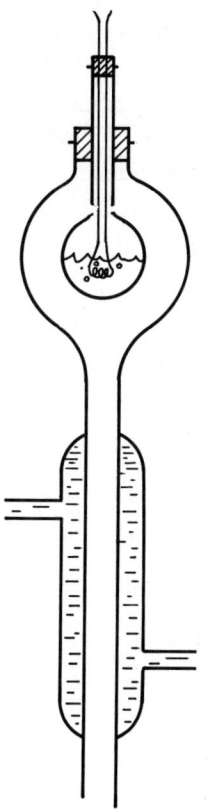

ELECTRICAL METHOD OF DETERMINING THE SPECIFIC LATENT HEAT OF EVAPORATION OF A LIQUID

Liquid is placed in the inner vessel, and when it boils the vapour passes around the liquid before being condensed so that it may be collected at the bottom of the apparatus. Since the boiling liquid is surrounded by its own vapour, heat losses from the inner vessel are reduced. When the apparatus has reached a steady state, with the liquid boiling freely, the mass of condensed liquid, m, is collected in a measured time, t, and weighed. Then:

$$IVt = ml + \text{heat losses}$$

where l is the specific latent heat of the liquid.

The heat losses depend on the temperature, which is fixed at the boiling-point of the liquid, and the time. The experiment is repeated by increasing the electrical power supplied, and the new mass collected in the same time is is measured. Then $I'V't = m'l + \text{heat losses}$. The heat losses are eliminated by subtraction, and l may then be calculated.

Figure 7

NEWTON'S LAW OF COOLING

Newton's law of cooling states that the rate of loss of heat of a body is proportional to the excess temperature over the surroundings, provided that the nature and area of the cooling surface remains constant. The law is true for excess temperatures only up to about 30 °C in still air, but for higher excess temperatures in a forced draught. A convenient way of obtaining a steady forced draught is to use a hair drier set to give a steady stream of cold air.

To verify the law, a calorimeter is filled with water which has been heated above room temperature. As the water cools, its temperature is recorded at minute intervals, and a cooling graph plotted of temperature against time. At points on this graph the rate of cooling is determined by drawing tangents to the curve and finding the slope of each. Thus $d\theta/dt$ is found for various values of θ. If θ_R is the room temperature, then $\theta - \theta_R$ is the excess temperature. A second graph is drawn of $d\theta/dt$ against $(\theta - \theta_R)$, and this graph should be a straight line through the origin, showing that the rate of cooling is proportional to the excess temperature. Since the heat capacity of the calorimeter and its contents is constant, it follows that the rate of loss of heat is also directly proportional to the excess temperature.

COMPARISON OF SPECIFIC HEAT CAPACITIES OF LIQUIDS BY THE METHOD OF COOLING

Equal volumes of water and liquid are allowed to cool as described above in identical calorimeters. The nature and area of the cooling surfaces must be the same in each case, which is why equal volumes are used. Cooling curves are drawn on the same axes for each set of measurements. A convenient temperature is chosen and the rate of cooling of each liquid is determined at this temperature. Since the cooling conditions in each case were identical, the rate of loss of heat is the same for both.

Hence $(m_1 c_1 + C)\left(\dfrac{d\theta}{dt}\right)_1 = (m_2 c_2 + C)\left(\dfrac{d\theta}{dt}\right)_2$

where m_1, m_2, c_1, c_2 are the masses of water and liquid, and specific heat capacities of water and liquid respectively, and C is the heat capacity of the calorimeter. Assuming c_1 to be $4\,200$ J kg^{-1}K^{-1}, a value for c_2 may be found.

This method does not really use Newton's law, although it is usually referred to as a cooling method, since only a single excess temperature is selected.

CORRECTION FOR HEAT LOSSES IN CALORIMETRY

Simple method of mixtures experiments are not sufficiently accurate at this level because of the inevitable heat losses which occur. If a hot object is placed in cold water, the resulting rise in temperature which may be measured is less than the theoretical value because of heat loss. The purpose of the correction is to compute the rise in temperature which would have occurred under ideal conditions of zero heat loss. A suitable experiment in which to apply the correction is in determining the specific heat capacity of a poor conductor, such as a rubber bung.

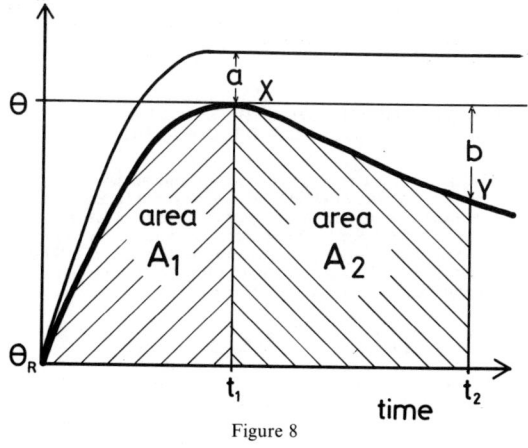

Figure 8

Instead of simply finding the highest temperature reached by the mixture, readings of the temperature are taken every half-minute, covering the period immediately following the mixing up to a point well past the time when the mixture has attained its highest temperature. A graph of excess temperature, $(\theta - \theta_R)$, against time, is drawn. The peak of the curve, X, is marked, and a line through X is drawn parallel to the time axis. A second point, Y, is selected. For accuracy, Y should be as far to the right as possible. Vertical lines through X and Y are drawn to cut the time axis at t_1 and t_2 respectively. By counting squares, or weighing, the areas A_1 and A_2 are found as accurately as possible. Assuming Newton's law:

$$\frac{dq}{dt} = k.\theta_E, \text{ where } \theta_E \text{ is the excess temperature } (\theta - \theta_R),$$ and k is a constant depending upon the nature and area of the calorimeter. For the period up to t_1:

$$q_1 = \int_0^{t_1} k.\theta_E \, dt = k \int_0^{t_1} \theta_E \, dt = kA_1$$

If this heat had not been lost, it would have warmed the calorimeter and contents by a degrees.

$\therefore Ca = kA_1$, where C is the heat capacity of the calorimeter and its contents.

Similarly, the heat lost between t_1 and t_2 is kA_2, and this loss caused the temperature of the calorimeter and contents to fall by b degrees.

$\therefore Cb = kA_2$

Hence $\dfrac{a}{b} = \dfrac{A_1}{A_2}$ or $a = b.\dfrac{A_1}{A_2}$

Once the temperature correction, a, has been calculated, the theoretical final temperature of the mixture is known, and the calculation of the specific heat capacity of rubber can proceed in the normal way.

The method is equally applicable to electrical methods of heating. The observed final temperature when the heater is switched off will be lower than the theoretical value due to heat losses. If temperature measurements continue after switching off, a graph similar to that shown above may be constructed. Note that the method of mixtures is a special case of the law of conservation of energy. Check any calculations that you may make to ensure that every term has been included.

THE KINETIC THEORY OF GASES

Four basic assumptions must be made:
1. **Collisions between molecules, and those between molecules and the walls of the vessel are perfectly elastic.**
2. **There is negligible attraction between the molecules.**
3. **The molecules occupy a negligible volume compared with the total volume of the gas.**
4. **The time between collisions is very much greater than the time during which a collision takes place.**

Consider a gas containing n molecules each of mass m, contained in a cubical box of side a. Three axes, x, y, and z, are parallel to the edges of the box. Any molecule with velocity c may be considered to have components of velocity u, v, and w parallel to these axes. When the molecule collides with one of the walls which is perpendicular to the x axis, the change in momentum is $mu - (-mu) = 2mu$. The time taken for the molecule to travel from this face to the opposite face and back is $2a/u$, and after this time the molecule makes another collision with the original face. Therefore the rate of change of momentum is $2mu \div 2a/u = mu^2/a$, and this is equal to the force exerted by the molecule on the wall.

The average value of the force exerted on this wall by all n molecules is $nm\overline{u^2}/a$, where $\overline{u^2}$ is the average value of u^2.

$$\overline{u^2} = \frac{u_1^2 + u_2^2 + u_3^2 + \cdots}{n}$$

The average pressure, P, on the wall = force/area = $nm\overline{u^2}/a^3$. But a^3 is the volume of the gas, V, so $PV = nm\overline{u^2}$.

For a large number of molecules moving with different speeds in random directions, the mean squares of the speeds parallel to all three axes must be the same, otherwise the pressure on different sides of the box would not be the same.

$$\overline{u^2} = \overline{v^2} = \overline{w^2}$$

But for each molecule, $c^2 = u^2 + v^2 + w^2$. Hence $c^2 = 3u^2$. Similarly, the mean square, $\overline{c^2}$, for all the molecules is given by $\overline{c^2} = 3\overline{u^2}$.

Hence $PV = \frac{1}{3}nm\overline{c^2}$

This analysis assumes the molecules do not collide with one another. They do, but the pressure is unaffected since the total momentum and kinetic energy of the gas are unchanged.

The equation $PV = \frac{1}{3}nm\overline{c^2}$ may also be written as $P = \frac{1}{3}\rho\overline{c^2}$, where ρ is the density of the gas, since nm is the total mass of all the molecules. In this form the equation enables the speed of molecules to be calculated. For example, the density of hydrogen is 0.09 kg m^{-3} at s.t.p. Under these conditions the pressure of the gas is $0.76 \times 13\,600 \times 9.81$ (height of mercury column $\times$ density of mercury $\times g$), so

$$\overline{c^2} = \frac{3P}{\rho} = \frac{3 \times 0.76 \times 13\,600 \times 9.81}{0.09} = 3.38 \times 10^6 \text{ m}^2 \text{ s}^{-2}$$

$\sqrt{\overline{c^2}}$ is called the **root mean square** speed of the molecules. In this case $c_{\text{r.m.s}} = \sqrt{3.37 \times 10^6} = 1.84 \times 10^3$ m s^{-1}.

The **mole** is the SI unit of amount of substance, being equal to the **amount of substance which contains as many elementary units as there are carbon atoms in 0.012 kg of carbon-12.** In the case of gases, the elementary units are referred to as molecules, and it follows from the above definition that **one mole of any gas will contain the same number of molecules.** Let this number be N_M. The kinetic theory equation developed on the previous page becomes $PV = \frac{1}{3}N_M\,m\overline{c^2}$ for 1 mole of gas. At the beginning of the section on heat, we referred to the ideal gas equation for 1 mole, $PV = RT$, where R is the molar gas constant. Since the left hand sides of these equations are the same:

$$RT = \frac{1}{3}N_M\,m\overline{c^2} = \frac{2}{3}N_M(\tfrac{1}{2}m\overline{c^2})$$

the $\frac{1}{2}m\overline{c^2}$ being the average kinetic energy of translation of the molecules ($\overline{KE}$).

Thus $\overline{KE} = \dfrac{3}{2}.\dfrac{R}{N_M}.T$. Since R and N_M are both constants, it follows that the average kinetic energy of translation of the molecules of an ideal gas is directly proportional to the absolute temperature of the gas.

It is important to specify translational kinetic energy, since gas molecules also have kinetic energy of rotation.

The ratio R/N_M is called the Boltzmann constant, k. Its value is 1.38×10^{-23} J K^{-1}. But $R = 8.314$ J mol^{-1}K^{-1} (page 22).

$$\therefore N_M = \frac{R}{k} = \frac{8.314}{1.38 \times 10^{-23}} = 6.02 \times 10^{23} \text{ mole}^{-1}$$

This number, the number of molecules in 1 mole, is known as **Avogadro's number.** Avogadro's original hypothesis was that **equal volumes of different gases contain the same number of molecules at the same temperature and pressure.**

1 mole of hydrogen has a mass of 0·002 016 kg. As previously stated, the density of hydrogen at s.t.p. is 0·09 kg m^{-3}. From these figures the volume of 1 mole of hydrogen is:

$$\frac{0·002\ 016}{0·09} = 0·0224 \text{ m}^{-3} \quad \text{or} \quad \textbf{22·4 litres}$$

and this is the **volume which 1 mole of any gas will occupy at s.t.p.**
It may be useful to consider the ideal gas equation in forms other than that used for 1 mole. In particular, if there are N moles of gas, the equation becomes $PV = NRT$. For unit mass of gas, the number of moles is $1/M$ where M is the mass of one mole of the gas. Therefore for m kg of gas:

$$PV = \frac{m}{M} RT$$

In the case of oxygen, for example, $M = 0·032$ kg, while for helium $M = 0·004$ kg.

DALTON'S LAW OF PARTIAL PRESSURES

Dalton's law of partial pressures states that **in a mixture of gases which do not react, the pressure exerted by the mixture is equal to the sum of the partial pressures of the constituents.** For example, in a mixture of 1 part oxygen to 4 parts nitrogen together exerting a pressure of 10 N m^{-2}, the pressure due to the oxygen is 2 N m^{-2} and that due to the nitrogen is 8 N m^{-2}.

KINETIC ENERGY AND TEMPERATURE

The equation $\overline{KE} = \dfrac{3}{2} \dfrac{R}{N_M} T = \frac{3}{2}kT$, where k is the Boltzmann constant,

indicates that for a mixture of gases, **where the temperatures of the constituents are the same, the average kinetic energies of the constituents are also the same.** If the molecules of one gas are heavier than those of the other, then $\overline{c^2}$ for the heavier gas will be smaller, but the $\overline{KE}_1$ will equal $\overline{KE}_2$, both being equal to $\frac{3}{2}kT$. This result may be used to derive theoretically **Graham's Law** which states that **the rate of diffusion of a gas is inversely proportional to the square root of its density.**

THE SPECIFIC HEATS OF GASES

When a gas is heated at constant volume, the heat supplied to it gives the molecules more kinetic energy. **The specific heat capacity at constant volume, c_v, is the heat required to raise the temperature of 1 kg of the gas by 1 K without a change in volume.**

If the gas is heated at constant pressure, more heat must be applied to an equal mass of gas to raise its temperature by the same amount, since, in this case, in addition to increasing the kinetic energy of the molecules, work must be done in pushing back the walls of the container. Thus if c_p represents the specific heat capacity of the gas at constant pressure, $c_p = c_v +$ **work done by gas in expanding.**

Consider 1 mole of a gas contained in a cylinder by a light, frictionless piston. $C_{p,m}$ and $C_{v,m}$ are respectively the molar heat capacities at constant pressure and constant volume. Let P be the pressure of the gas and V_1 its initial volume. Let the piston have a cross-sectional area of A. If the gas is heated through 1 K, with the pressure remaining constant, let dx be the distance which the piston is pushed back, so that the volume of the gas increases to V_2.

Work done by gas in expanding = force × distance = $PA dx$.

But $A dx$ is the change in volume of the gas, i.e. $V_2 - V_1$.

∴ work done by gas in expanding = $P(V_2 - V_1)$.

By Charles' law, $\dfrac{V_1}{T} = \dfrac{V_2}{T+1}$ where T is the initial temperature of the gas which is then raised by 1 K.

Hence $V_2 - V_1 = V_1/T$ and:

work done in expanding gas = $P(V_2 - V_1) = PV_1/T = R$.

∴ $C_{p,m} = C_{v,m} + R$

For unit mass of gas, $c_p = c_v + r$, where r is the gas constant for unit mass. By the result on the previous page, $r = R/M$. A gas may have an infinite number of heat capacities, since there is no reason why either the pressure or the volume should remain constant. The two heat capacities previously referred to are known as the principal heat capacities. c_p is the maximum, and c_v the minimum. All other heat capacities lie between these extremes, the exact value depending on the conditions under which the gas is heated.

In deriving the equation $PV = \frac{1}{3}nm\overline{c^2}$, we resolved the velocity of a molecule, c, into three mutually perpendicular components, u, v, and w. The argument was made that $\overline{u^2} = \overline{v^2} = \overline{w^2}$, since the pressure on all walls is the same. $\therefore \overline{u^2} = \overline{v^2} = \overline{w^2} = \frac{1}{3}\overline{c^2}$. The average kinetic energy of a molecule, $KE = \frac{1}{2}m\overline{c^2} = kT$. The molecule is said to have three **degrees of freedom,** each one corresponding to one of the axes along which its velocity is resolved. Thus the average kinetic energy of translation in each degree of freedom is:

$$\frac{1}{2}m\overline{u^2} = \frac{1}{2}m\overline{v^2} = \frac{1}{2}m\overline{w^2} = \frac{1}{2}kT$$

and the molecule has kinetic energy $\frac{1}{2}kT$ per degree of freedom. In practice the kinetic energy of a molecule may be rotational as well as translational. A monatomic molecule has negligible kinetic energy of rotation. A diatomic molecule has negligible rotational K.E. if it is spinning about the axis through the two atoms. Any other rotational K.E. may be resolved into two components at right angles to this axis, and the molecule is said to have 2 degrees of rotational freedom. A general polyatomic molecule, where the atoms are not in line, has 3 degrees of rotational freedom.

Maxwell assumed that the average kinetic energy of a molecule was $\frac{1}{2}kT$ for each degree of freedom, rotational as well as translational. This assumption, called the **principle of equipartition of energy,** is true except at very low temperatures. Therefore KE of monatomic molecule $= \frac{3}{2}kT$ (translation only)

KE of diatomic molecule $= \frac{3}{2}kT$ (trans.) $+ \frac{2}{2}kT$ (rot.) $= \frac{5}{2}kT$

KE of polyatomic molecule $= \frac{3}{2}kT$ (trans.) $+ \frac{3}{2}kT$ (rot.) $= \frac{6}{2}kT$.

Consider unit mass of gas containing n molecules. If the gas is monatomic, the total KE of its molecules is $\frac{3}{2}nkT = \frac{3}{2}rT$, where r is the gas constant for unit mass. To raise the temperature by 1 degree, additional energy $\frac{3}{2}r$ must be supplied. But this is the specific heat capacity at constant volume, c_v.

$\therefore c_p = c_v + r = \frac{3}{2}r + r = \frac{5}{2}r$

The **ratio of the principal specific heat capacities,** γ is given by $\dfrac{c_p}{c_v} = \frac{5}{2}r \div \frac{3}{2}r = \frac{5}{3} = 1 \cdot 67$

For diatomic gases: $\dfrac{c_p}{c_v} = \frac{7}{2}r \div \frac{5}{2}r = \frac{7}{5} = 1 \cdot 4$

For polyatomic gases: $\dfrac{c_p}{c_v} = \frac{8}{2}r \div \frac{6}{2}r = \frac{8}{6} = 1 \cdot 33$.

ISOTHERMAL AND ADIABATIC CHANGES

If the temperature of a fixed mass of an ideal gas remains constant, then PV = constant (Boyle's Law). A graph of pressure against volume is a rectangular hyperbola, the curve being known as an **isothermal.** When a gas is compressed, or allowed to expand, under conditions of constant temperature, the resulting changes in pressure and volume are referred to as **isothermal compressions or expansions.**

A change in the total energy must accompany an isothermal. In the case of an expansion, energy must be supplied, and the work done is given by

$$q = \int_{V_1}^{V_2} PdV = rT \log_e \left(\frac{V_2}{V_1} \right)$$

where V_1 and V_2 are the initial and final volumes of the gas. An **adiabatic change** is one in which **no heat enters or leaves the system.** In the case of an adiabatic expansion, the **energy needed to expand the gas must therefore come from the gas itself,** and as a result of this work being done, the energy of the gas, and hence its temperature, must fall. For adiabatic changes, the relationship between pressure and volume is given by PV^{γ} = constant, where γ is the ratio of the principal specific heat capacities of the gas.

CLEMENT AND DESORMES' METHOD FOR γ

A large vessel contains a little concentrated sulphuric acid to dry any gas in it. A tightly fitting bung has 3 outlets, one connected to a pump, one to a manometer containing light oil, and the third to a tap. Air is pumped into the vessel, raising the pressure above atmospheric. The pressure is read from the manometer. The tap is opened briefly and then closed. When the manometer has settled, the pressure is again read. The vessel must be well lagged, and the assumption then is that the gas expands adiabatically when the tap is opened. The pressure falls rapidly, but then rises as the temperature of the gas increases once more when the tap has been closed. γ is calculated as $h_1/h_1 - h_2$, where h_1 and h_2 respectively represent the pressures before and after opening the tap. A series of results should be obtained.

CRITICAL TEMPERATURE

Though normally a gas at room temperature, ammonia may readily be liquefied by compressing it. Oxygen, however, may not be liquefied at room temperature, no matter how greatly it is compressed. Every gas has a **critical temperature.** For ammonia it is 132°C, while for oxygen it is −119°C. **Above its critical temperature** a substance may not be liquefied by pressure alone, and is called a **gas. Below the critical temperature** the substance may be liquefied by compression and is called a **vapour.** The **critical pressure** is the pressure which is just sufficient to liquefy a vapour at its critical temperature. The further the temperature is reduced, the less excess pressure will be required to bring about liquefaction, until, at its boiling point, the substance liquefies at standard atmospheric pressure.

VAN DER WAALS' EQUATION

An **ideal gas** is one which would completely obey the gas laws at all temperatures and pressures. A **real gas** does not obey the gas laws at all temperatures and pressures. Best agreement is found with the so called "permanent" gases, i.e. those with low critical temperatures. The four basic assumptions which were made at the beginning of the section on kinetic theory are true at normal temperatures and pressures, but become less realistic under extreme conditions. If the space through which the molecules are not free to move is b, then the volume in which the molecules are free to move is $(V - b)$. It should be noted that b is not the actual volume of the molecules, since there is a minimum distance of approach.

In a real gas, the attraction between the molecules cannot be neglected. A molecule about to hit the wall of the vessel will be attracted by other molecules near the wall, and will strike with a lower velocity than might have been expected. Thus the true pressure within the bulk of the gas will be less than the observed pressure. Van der Waals showed that the correction necessary is inversely proportional to the square of the volume. Applying these two corrections, we have, for a real gas:

$$\left(P + \frac{a}{V^2}\right)(V - b) = RT$$

VAPOURS

A **saturated vapour** is one which is in contact with its own liquid in an enclosed space. **Dynamic equilibrium** means that the same number of molecules leave the liquid as enter it. The pressure exerted by a saturated vapour **(S.V.P.)** is independent of the volume, because when the space above the liquid is increased, more molecules leave the liquid than return until a new equilibrium is established. When the space is reduced, vapour molecules condense. S.V.P. increases with temperature. An unsaturated vapour obeys the gas laws approximately, provided the vapour content is not too great. **A liquid boils when its S.V.P. is equal to the external pressure** acting on it, and this enables the variation of S.V.P. with temperature to be established.

The liquid is boiled and the temperature noted. Condensed vapour returns to the flask. The large central flask minimises the effects of sudden pressure changes. The manometer records the pressure inside the apparatus, and the pressure over the boiling liquid is atmospheric ± the manometer reading. This pressure is equal to the S.V.P. at the indicated temperature. The pressure may be varied by means of the pump, and values of S.V.P. found at temperatures either side of the boiling point.

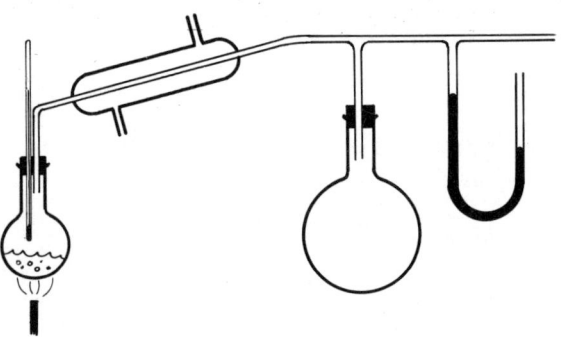

Figure 9

THERMAL CONDUCTIVITY

In the steady state, the quantity of heat flowing along a bar of cross-sectional area A and length l in time t is given by $Q = kAt\dfrac{(\theta_2 - \theta_1)}{l}$, where θ_2 and θ_1 are the temperatures at the hot and cold ends of the bar and k is the thermal conductivity. The ratio $(\theta_2 - \theta_1)/l$ is called the **temperature gradient.** If this, together with the time and cross-sectional area, is unity, then $Q = k$, so the thermal conductivity may be defined as the quantity of heat per second flowing in the steady state between opposite faces of a unit cube of the material, these opposite faces being maintained at a temperature difference of 1 K, and there being no heat lost from the sides of the cube. The temperature gradient down a well-lagged bar is constant, but if the bar is not lagged heat is lost from the sides of the bar. The rate of loss of heat will be greater at the hot end of the bar, so the temperature gradient will be steeper at this end than at the colder end.

DETERMINATION OF k FOR A GOOD CONDUCTOR

The apparatus consists of a metal bar, about 0·4 m long and 0·05 m in diameter. One end is fitted with a steam chest, and the other has a copper spiral wound round it, through which water is passed. The temperatures of the water as it enters and leaves the spiral, (θ_3 and θ_4) may be measured. The temperature gradient along the bar can be found from the readings of the two thermometers placed along the bar,

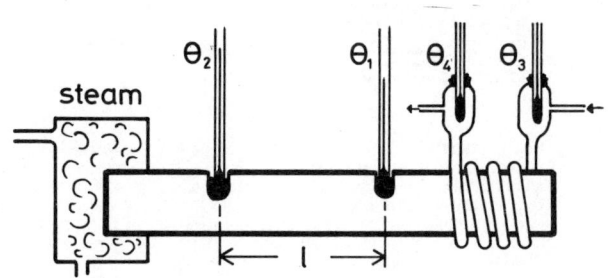

Figure 10

(θ_1 and θ_2), which are separated by a distance l. The cross-sectional area, A, is calculated by measuring the diameter of the bar with callipers. Heat losses are minimised by thoroughly lagging the whole apparatus. Steam is passed into the chest, and water from a constant head device is passed through the copper spiral. The apparatus is left until all temperatures have become steady. The mass of water, m, flowing through the spiral in time t is measured, and the four temperatures noted. Then:

$$mc(\theta_4 - \theta_3) = kAt(\theta_2 - \theta_1)/l$$

where c is the specific heat capacity of water.

The water must flow in the direction indicated, so as not to reverse the temperature gradient along the bar. The thermometers along the bar are inserted into shallow holes drilled into the bar, and a little mercury is used to ensure good thermal contact. The thermometers should be interchanged to overcome errors due to their inaccuracy.

DETERMINATION OF k FOR A BAD CONDUCTOR

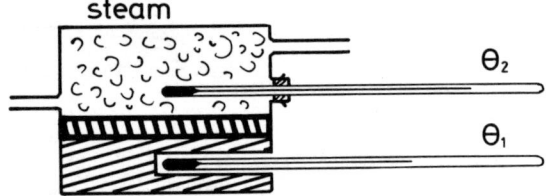

Figure 11

The bad conductor is in the form of a disc about 2 mm thick. The thickness, l, and cross-sectional area, A are measured. A heavy brass slab is suspended from a stand, the specimen disc placed on it, and on top of that is placed a steam chest. Both the slab and the steam chest are tin-plated to reduce heat loss, and have holes drilled in them to take thermometers. Steam is passed into the chest and the apparatus left until the two thermometer readings become steady. Then:

$$\frac{Q}{t} = kA \frac{(\theta_2 - \theta_1)}{l}$$

To find Q/t, the specimen is removed and the steam chest placed in direct contact with the lower slab, heating the latter to a temperature well above θ_1. The chamber is then removed and the disc replaced. The slab is allowed to cool and a cooling graph drawn by measuring the temperature at minute or half-minute intervals. The rate of cooling at θ_1 is found by drawing a tangent to the graph and measuring the slope. The rate of loss of heat is then $mcd\theta/dt$, where m and c are the mass and specific heat capacity of the slab and $d\theta/dt$ is the rate of cooling at θ_1. The assumption is that the determined rate of loss of heat of the slab is the same as that which it experienced during the first part of the experiment. Since the temperatures were steady, this must have been equal to the rate at which heat was passing through the specimen.

It is important that the slab be allowed to cool under conditions identical with those which existed during the first part of the experiment. In that situation it was able to lose heat only from the base and sides, which is why the poorly conducting specimen is placed on top for the second part of the experiment. The tinned surfaces must be clean, and a smear of vaseline used to improve thermal contact between the metal and the specimen disc.

DETERMINATION OF k FOR A TUBE

The tube is mounted in a wider glass tube through which steam is passed. Water from a constant head device is passed through the specimen tube, which is inclined slightly to prevent air bubbles forming

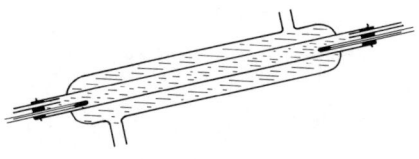

Figure 12

in it. Two thermometers record the temperature of the water (θ_1 and θ_2) as it enters and leaves the specimen tube. The rate of flow of water should be adjusted so that θ_2 is about 10 degrees higher than θ_1. When the temperatures are steady, these temperatures are noted, and the rate of flow of water is measured. If a mass m is collected in time t, then the heat flowing per second through the walls of the tube is $mc(\theta_2 - \theta_1)/t$. If r_1 and r_2 are respectively the internal and external radii of the tube, and if its length is l, then $(r_2 - r_1)$ is the thickness of the tube and $2\pi \times \frac{1}{2}(r_1 + r_2) \times l$ is its mean area. If θ_3 is the temperature of the steam, the temperature drop across the walls of the tube is $(\theta_3 - \frac{1}{2}(\theta_1 + \theta_2))$.

Thus k may be found from the expression:

$$\frac{mc(\theta_2 - \theta_1)}{t} = \frac{k \cdot 2\pi \times \frac{1}{2}(r_1 + r_2) \times l \times (\theta_3 - \frac{1}{2}(\theta_1 + \theta_2))}{r_2 - r_1}$$

CONDUCTIVITY THROUGH COMPOSITE MATERIALS

Problems involving the passage of heat through, say, a lagged copper boiler, may often be solved by considering the copper and the lagging to be together equivalent to a single thickness of copper. Thus if the copper is 5 mm thick, and the lagging is 20 mm thick, and if the thermal conductivities are respectively 380 W m^{-1} K^{-1} and 0·038 W m^{-1} K^{-1}, then it will be noted that these conductivities are in the ratio 10 000 to 1. Thus 1 mm of lagging is equivalent to 10 000 mm of copper and the combination is equivalent to 200 000 + 5 mm of copper.

If the temperature of the interface is required, knowing the inside and outside temperatures, then two simultaneous equations must be constructed. For example, in the case of doubly glazed windows, if the room temperature is 20°C and the outside temperature 0°C, and if θ_2 and θ_1 respectively are the temperatures at the points where the air cushion is in contact with the glass, then:

$$\frac{k_g(20 - \theta_2)}{l_g} = \frac{k_a(\theta_2 - \theta_1)}{l_a} = \frac{k_g(\theta_1 - 0)}{l_g}$$

where l_g and l_a are the thicknesses of glass and air, and k_g and k_a their thermal conductivities. From these equations θ_2 and θ_1 may be calculated.

RADIATION

Infra-red heat radiation travels with the speed of light. It has similar properties of reflection and refraction, but the wavelength is less than that of light. Infra-red radiation falling on matter is absorbed, thereby raising the kinetic energy of the molecules.

All substances radiate heat at all temperatures above absolute zero. It is only at comparatively high temperatures that the radiated heat is significant.

A **Black Body** is one which absorbs completely all radiation falling upon it. When heated a black body therefore emits radiation of all wavelengths.

Stefan's Law states that the total radiation emitted in unit time from unit area of a black body is directly proportional to the fourth power of its absolute temperature. Remembering that the black body will also be receiving energy from its surroundings, Stefan's Law may be written as $E = \sigma.A.t \ (T_2^4 - T_1^4)$ where E is the net energy radiated in time t by a black body whose radiating surface has an area A m^2. T_2 is the temperature of the black body and T_1 the temperature of the surroundings which receive the radiation. If E is measured in watts, then σ has the units of watt. m^{-2}.$^{\circ}$K^{-4}.

Newton's Law of Cooling is a special case of Stefan's Law. When the temperature difference between the body and its surroundings is small, $(T_2^4 - T_1^4)$ is approximately proportional to $(T_2 - T_1)$.

DISTRIBUTION OF ENERGY IN THE SPECTRUM OF A BLACK BODY

The graph shows the intensity of energy plotted against wavelength. As the temperature rises so the peaks of the curves gradually shift towards the visible spectrum, hence the colour of a hot body is characteristic of its temperature.

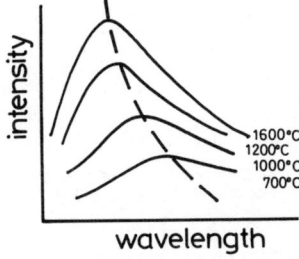

Figure 13

Geometrical Optics

CURVED MIRRORS

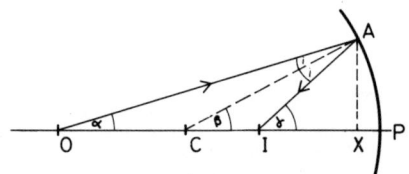

Figure 14

In the diagram O is a point object on the **principal axis** of a concave spherical mirror. The incident ray, OA, forms an angle of incidence i with the normal, CA, where C is the **centre of curvature**. The reflected ray is AI, in accordance with the laws of reflection. A second ray, OP, passes along the principal axis, so that a point image, I, is formed. The image is real. α, β, and γ are the angles indicated on the diagram.

From geometry, it will be seen that $\beta = \alpha + i$, and $\gamma = \beta + i$. Hence $i = \beta - \alpha = \gamma - \beta \therefore \alpha + \gamma = 2\beta$

If the angles α, β, and γ are very small, we may assume that:

1. X is coincident with P,
2. the tangents of the angles are equal to the angles.

It is also advisable to include a sign for each angle at this stage. In this case all three angles are positive. Hence:.

$$\frac{AX}{OP} + \frac{AX}{IP} = 2\frac{AX}{CP} \quad \text{or} \quad \frac{1}{u} + \frac{1}{v} = \frac{2}{r}$$

In the special case of light parallel to the principal axis striking the mirror, the image is formed at the **principal focus**. In that case:

$$\frac{1}{\infty} + \frac{1}{f} = \frac{2}{r}, \qquad \therefore \quad \frac{1}{u} + \frac{1}{v} = \frac{1}{f}$$

It is most important to use a **sign convention,** which covers all combinations of real and virtual images, as well as both convex and concave mirrors. The **real is positive convention** gives a positive sign to real objects and images, and a negative sign to virtual objects and images. For concave mirrors, f and r are positive; for convex mirrors they are negative. The **linear magnification** is v/u, and this may be combined with the formula $\dfrac{1}{u} + \dfrac{1}{v} = \dfrac{1}{f}$ to give $m = \dfrac{v}{f} - 1$. This latter is often a more convenient way of calculating magnification.

DETERMINATION OF FOCAL LENGTH AND RADIUS OF CURVATURE FOR CURVED MIRRORS

If $u > f$, a concave mirror gives a real image, so determination of the focal length, and hence the radius of curvature, presents no problems. Using either an illuminated object and screen, or pins and no-parallax, a series of values of u and corresponding v's may be found. A graph of $1/u$ against $1/v$ will give a straight line of negative slope which intercepts both axes at $1/f$. In the special case $u = 2f$, the image is formed coincident with the object. By adjusting u until object and image are alongside each other and the same size, the radius of curvature may quickly be found, but only one value is obtained in this way.

For convex mirrors, the image is normally virtual, so that a lamp and screen may not be used. Pins and no parallax are possible, but a better method is to use a virtual object in order to obtain a real image. An illuminated object is set up in front of a converging lens, and a screen moved to receive the sharpest image possible. The position of the screen is noted, C, and the mirror inserted between the lens and the screen. Without moving anything else, the mirror's position is adjusted until image and object coincide at O. In this case the light must strike the mirror normally, since it is reflected back along its incident path, i.e. the mirror has been placed so that its centre of curvature is at C. If M is the position of the mirror, the distance MC is equal to its radius of curvature. The experiment may be repeated to obtain a series of values of the radius of curvature, and hence the focal length.

REFRACTION AT PLANE SURFACES

When light passes from medium 1 into medium 2, the **refractive index** $_1n_2$ is defined as the ratio:

$$\frac{\text{velocity of light in medium 1}}{\text{velocity of light in medium 2}}$$

Also $_1n_2 = \sin i_1/\sin i_2$. For the **absolute refractive index,** medium 1 is a vacuum. The frequency of the radiation is not changed at the boundary; the wavelength is reduced when passing into the medium of higher absolute refractive index.

When light travels from a medium of higher absolute refractive index, the angle of refraction exceeds the angle of incidence. Total internal reflection occurs when the angle of incidence is greater than the critical angle. The latter is the angle of incidence for which the emerging ray has $i_2 = 90°$.

Real and Apparent Depth method for n.

A mark is made on the inside surface of a hollow tank. A travelling microscope is racked up and down until it is sharply focused on the mark, and the reading on the vertical scale noted. Liquid is poured into the tank and the microscope refocused on the mark. A second reading is taken. Finally some lycopodium powder or other fine material is sprinkled on the top of the liquid and, after refocusing, a third reading is taken. The real depth of the liquid is $1 - 3$, and the apparent depth of the image is $2 - 3$. The refractive index of the liquid relative to air is the real depth divided by the apparent depth. A series of values may be obtained by varying the depth of the liquid. The method may also be used using a solid transparent block with parallel sides, although it may be possible to obtain only one set of readings.

The Air Cell.

The air cell consists of a thin film of air enclosed between two glass plates. It may be rotated about a vertical axis, and a pointer and scale are incorporated to measure this rotation. The cell is immersed in the liquid which is contained in a parallel-sided tank, and illuminated with monochromatic light. A small eyepiece is used for viewing, and usually this has a vertical crosswire. The air cell is rotated until half the field of view is dark. This occurs when the light at the glass-air boundary reaches the critical angle. The cell is then rotated to locate the symmetrical

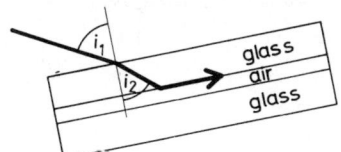

Figure 15

position in the opposite inclination, when the other half of the field of view is dark. Taking two readings like this obviates the necessity of aligning the cell so that it is initially perpendicular to the incident light.

From the above diagram, $\dfrac{\sin i_1}{\sin i_2} = {_L}n_G$, and $\dfrac{\sin i_2}{\sin 90} = {_G}n_A$

Hence ${_L}n_A = {_L}n_G \times {_G}n_A = \dfrac{\sin i_1}{\sin i_2} \times \dfrac{\sin i_2}{\sin 90} = \sin i_1$

$\therefore {_A}n_L = \dfrac{1}{{_L}n_A} = \dfrac{1}{\sin i_1}$

In the method described above, $2i_1$ is the total angle through which the cell is rotated between the two positions of extinction. The liquid must, of course, have a lower refractive index than the glass, otherwise extinction will occur due to internal reflection at the liquid-glass boundary.

REFRACTION THROUGH PRISMS

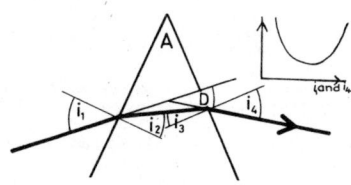

Figure 16

45

The diagram on the previous page shows the path of a typical ray of monochromatic light through a triangular prism. Using a light box, a series of values of i_1 and i_4 could be found, and in each case the angle of deviation, D, could be measured. If a graph of D against i_1 and i_4 were drawn, a curve similar to that shown would be obtained. This curve indicates that there is a minimum angle of deviation for the prism, and that at this minimum deviation $i_1 = i_4$. Let the angle of minimum deviation be D_{min}.

Using the exterior angle property of triangles, and noting that the quadrilateral formed by the sides of the prism and the normals is cyclic:

$$A = i_2 + i_3 \qquad \text{and} \qquad D = (i_1 - i_2) + (i_4 - i_3)$$

At minimum deviation, $i_1 = i_4$, and therefore $i_2 = i_3$, so:

$$i_2 = A/2 \qquad \text{and} \qquad i_1 = (A + D_{min})/2$$

The refractive index of the material of the prism is given by:

$$n = \frac{\sin i_1}{\sin i_2} = \sin \frac{(A + D_{min})}{2} \bigg/ \sin \frac{A}{2}$$

THE SPECTROMETER

As its name implies, the spectrometer is principally used for investigating spectra, but it may also be used to determine the refractive index of the material of a prism, using the theory outlined above. The spectrometer consists essentially of a collimator, whose function is to provide a narrow beam of parallel light, and a telescope which brings the light to a focus on a pair of mutually perpendicular crosswires. Between the collimator and the telescope is a table which may be rotated. The telescope may also be rotated, and both table and telescope have scales with verniers.

Before use the spectrometer must be adjusted. This involves:

1. Adjusting the telescope to receive parallel light by looking through it at a distant object and focusing so that there is no parallax between the image and the crosswires.
2. Adjusting the collimator to send out parallel light by focusing it so that there is no parallax between the image of the slit and the crosswires.
3. Adjusting the prism table to be horizontal. This must be done by an optical method.

DETERMINATION OF n FOR A PRISM

Finding A. The prism is set up with its refracting angle pointing towards the collimator. The telescope is rotated until it receives the image of the slit by reflection from one face. The scale reading is noted and the telescope turned to receive the image by reflection from the other face. The angle between these two positions is 2A. Two positions are used since it is not then necessary to place the prism symmetrically on the table.

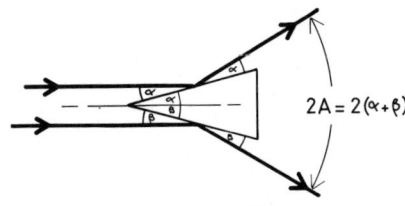

$$2A = 2(\alpha + \beta)$$

Figure 17

Finding D_{min}. With the prism placed as shown, the table is rotated anti-clockwise, and the image of the slit kept at the centre of the crosswires by rotating the telescope at the same rate. As angle D gets smaller, there comes a point at which continued rotation of the table causes the image of the slit to retreat from the line AB—it is impossible to obtain an image between the lines AB and CD, and CD corresponds to minimum deviation. The prism is rotated through 180° and the procedure repeated. The angle between CD and its corresponding position on the other side of AB is therefore equal to $2D_{min}$.

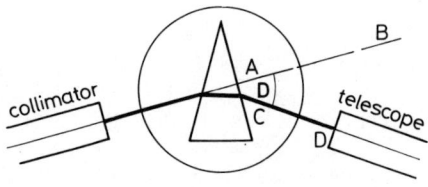

Figure 18

PRISMS WITH SMALL ANGLES

If the angles of incidence and refraction are small, then since $\sin \theta \doteqdot \theta$ under these circumstances, $i_1 = ni_2$ to a fair approximation.

The deviation, D, of a ray passing through a prism is given by $D = (i_1 - i_2) + (i_4 - i_3)$ (page 45). If the refracting angle, A, of the prism is also small, then since $A = i_2 + i_3$ and both A and i_2 are small, it follows that i_3 is also small. The angle of emergence, i_4, is small, so $i_4 = ni_3$.

$$\therefore D = (i_1 - i_2) + (i_4 - i_3) = (ni_2 - i_2) + (ni_3 - i_3)$$
$$= i_2(n - 1) + i_3(n - 1) = (n - 1)(i_2 + i_3) = (n - 1).A$$

Note that the deviation in this case is independent of i_1.

DISPERSION

The diagram shows white light incident at a small angle on a prism with small refracting angle. The light is deviated, but also dispersed into a spectrum of which three parts, red, yellow and blue are shown.

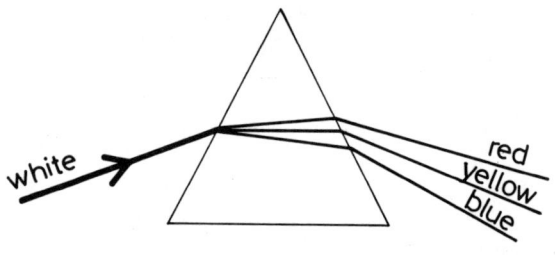

Figure 19

Hence $D_B = A(n_B - 1)$ and $D_R = A(n_R - 1)$

The angular dispersion is equal to $(D_B - D_R)$ and is equal to $A(n_B - 1) - A(n_R - 1)$.

Hence $(D_B - D_R) = A(n_B - n_R)$.

The mean deviation of the light is often equated to the deviation of the yellow light, since this colour lies roughly in the centre of the spectrum.

The dispersive power of the material of the prism is defined as:

$$= \frac{\text{angular dispersion between blue and red (extreme) rays}}{\text{mean deviation (deviation of yellow ray)}}$$

Hence $= \dfrac{(D_B - D_R)}{D_Y} = \dfrac{A(n_B - n_R)}{A(n_Y - 1)} = \dfrac{(n_B - n_R)}{(n_Y - 1)}$

ACHROMATIC COMBINATION OF PRISMS

Dispersion in prisms is often a nuisance, and may be overcome by placing a second prism, inverted, in contact with the first. If the second prism has a different refracting angle and dispersive power from the first one, it is possible to avoid dispersion without preventing deviation. The condition for the two prisms to form **an achromatic doublet** is that the angular dispersions of the pair must be equal. Hence:

$$A(n_B - n_R) = A^1(n_B^1 - n_R^1)$$

SPECTRA

1. Line spectra. If a sodium source is viewed using a prism and spectrometer, the light observed consists of two yellow lines close together. Other elements also give characteristic lines, the wavelengths of these lines depending on the type of atoms present. Measurement of wavelength, using a diffraction grating instead of a prism, enables elements to be identified. The wavelengths of the two sodium lines are 589·0 and 589·6 nm. The origin of line spectra is discussed more fully in the section on atomic structure.

2. Continuous spectra. If a white light source, such as an electric lamp filament, is viewed, a continuous spectrum is seen. This is a complete range of colours, from red to violet, which merge into one another. In general, continuous spectra are obtained from liquids and solids.

3. Band spectra. These are spectra made up of groups of very many lines close together, each band appearing as a short section of a continuous spectrum. A spectrometer and diffraction grating of high resolving power are needed to resolve the fine lines. Unlike line spectra, band spectra are characteristic of molecules rather than atoms. Examples of band spectra are those obtained with nitrogen and oxygen.

Spectra may be classified according to their method of production. The four main types are flame spectra, spark spectra, arc spectra and discharge-tube spectra.

Absorption spectra. If a sodium flame is placed between the slit of a spectrometer and a source of continuous spectrum such as an arc lamp, the spectrum is crossed by two dark lines very close together. If the continuous spectrum is removed, it is seen that the two yellow sodium lines referred to above coincide with these two dark lines. The sodium flame has absorbed the light whose wavelengths correspond to its two lines. It radiates the yellow light again, but in all directions so that the intensity in the direction of the line joining the two sources is much reduced.

The sun gives a continuous spectrum which includes the infra-red and ultra-violet radiation. The visible spectrum is crossed by a number of dark **Fraunhofer lines.** These are due to the absorption of the spectral lines from various elements in the sun's atmosphere, the absorption taking place in the cooler layers of gas which surround the sun.

REFRACTION AT CURVED SURFACES

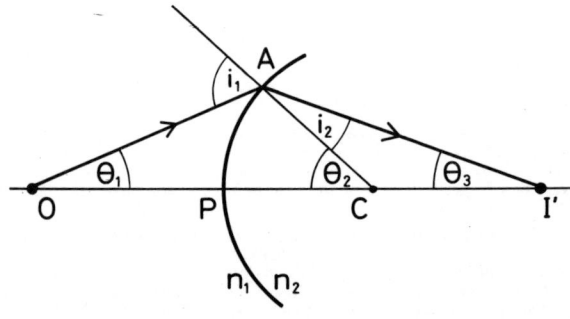

Figure 20

In the diagram on the previous page, O is a point object on the principal axis OP, a distance u from P and situated in a medium whose absolute refractive index is n_1. OA is an incident ray, and AI' the corresponding refracted ray A point image, I', is formed in the second medium whose absolute refractive index is n_2, and PI' is equal to v'. C is the centre of curvature, and $PC = r_1$.
$i_1 = \theta_2 + \theta_1$ and $i_2 = \theta_2 - \theta_3$. If the angles of incidence and refraction are small, then $i_1/i_2 = n_2/n_1$, so:

$$n_2(\theta_2 - \theta_3) = n_1(\theta_2 + \theta_1)$$

If the angles θ_1, θ_2, and θ_3 are very small, we may assume that:
1. AP is perpendicular to the principal axis,
2. the tangents of the angles are equal to the angles.

Hence:
$$\frac{n_2 \, AP}{r_1} - \frac{n_2 \, AP}{v'} = \frac{n_1 \, AP}{u} + \frac{n_1 \, AP}{r_1}$$

i.e.
$$\frac{n_1}{u} + \frac{n_2}{v'} = \frac{n_2 - n_1}{r_1}$$

It is necessary to specify a **sign convention,** and, as with curved mirrors, the **real is positive** convention makes all real object and image distances positive, while those for virtual objects and images are negative. If a surface is convex to the less optically dense medium, the light will be converged, and the surface is said to have positive radius. If the surface is concave to the less dense medium, the light will be diverged and the radius is negative. The value of $(n_2 - n_1)/r$ is called the power of the surface. Since refractive index is defined as a ratio of velocities, it has no sign, and the quantity $(n_2 - n_1)$ is always positive, representing in all cases the smaller refractive index subtracted from the larger one.

THIN LENSES

Starting with the formula $\dfrac{n_1}{u} + \dfrac{n_2}{v'} = \dfrac{n_2 - n_1}{r_1}$, we now consider the effect of a second refracting surface placed very close to the first, the pair forming a thin lens. I' acts as an object for this second surface, and it is a virtual object since it lies behind the surface. Suppose the radius of curvature of the second surface is r_2, and that it is convex to the less

51

dense medium. A final, real image will be formed at a distance v from the surface. Then:

$$-\frac{n_2}{v'} + \frac{n_1}{v} = \frac{n_2 - n_1}{r_2}$$

If the lens is thin enough for the two surfaces to be considered coincident, the two equations may be added:

$$\frac{n_1}{u} + \frac{n_1}{v} = (n_2 - n_1)\left(\frac{1}{r_1} + \frac{1}{r_2}\right) \qquad \text{(i)}$$

When rays of light close and parallel to the principal axis are refracted by a lens, they converge to a single point on the principal axis. The point is termed the principal focus, and is distance f (the focal length) from the lens. Hence:

$$\frac{n_1}{\infty} + \frac{n_1}{f} = (n_2 - n_1)\left(\frac{1}{r_1} + \frac{1}{r_2}\right) \qquad \text{(ii)}$$

From equations (i) and (ii):

$$\frac{n_1}{u} + \frac{n_1}{v} = \frac{n_1}{f} \qquad \text{or} \qquad \frac{1}{u} + \frac{1}{v} = \frac{1}{f}$$

With the real is positive convention, f is positive for a converging lens, and negative for a diverging lens.

The power of a lens is the reciprocal of its focal length in metres. Thus a converging lens of focal length 20 cm has a power of $+5$D. A diverging lens of focal length 40 cm has a power of $-2\frac{1}{2}$D.

REAL AND VIRTUAL IMAGES

A converging lens produces a real image if $u > f$, and a virtual image if $u < f$. If u lies between f and 2f, the real image is magnified, while for $u > 2f$, the real image is diminished. A diverging lens always produces a virtual image if the object is real, but can be used to give a real image of a virtual object. Suppose a converging lens produces a real image such that the total distance between object and image is d. If u is the object distance, then $v = (d - u)$. Substituting in the lens equation gives:

$$\frac{1}{u} + \frac{1}{d-u} = \frac{1}{f} \qquad \therefore \quad \frac{1}{f} = \frac{d}{u(d-u)}$$

Therefore $u^2 - du + df = 0$ and for real values of u, $d^2 > 4df$, or $d > 4f$.

TWO THIN LENSES IN CONTACT

By a method analogous to that by which the lens formula was derived, a formula for two thin lenses in contact may be established. The first lens, of focal length f_1 produces an intermediate image which acts as a virtual object for the second lens of focal length f_2. If F is the focal length of the combination: $\dfrac{1}{F} = \dfrac{1}{f_1} + \dfrac{1}{f_2}$.

DETERMINATION OF FOCAL LENGTH

For a converging lens the displacement method may be used. Object and screen are set up further apart than $4f$, f having been roughly determined by focussing light from a distant object onto a screen. From the equation $u^2 - du + df = O$, there will be two positions of the lens for which a real image may be obtained on the screen. Both these positions are located, and their separation, l, is noted. In the one case the object distance is u and the image distance v; in the other case the object distance is v and the image distance u.

$d = (u + v)$ and $l = (d - 2u)$.

Hence $u = (d - l)/2$ and $v = (d - u) = (d + l)/2$.

Hence $f = \dfrac{d^2 - l^2}{4d}$.

The main advantage of the method is that it is not necessary to measure to the lens itself—the displacement of any fixed point on the lens mounting may be used to determine l. It is worth noting that magnification = v/u in one case and u/v in the other. Hence the product of the magnifications produced in each case is 1, and their quotient is v^2/u^2.

For a diverging lens, the best method is first to set up an illuminated object and screen with a converging lens between them so that a real image is produced on the screen. The diverging lens is then inserted between the converging lens and the screen. The image produced by the converging lens acts as a virtual object for the diverging lens and a real image is found by moving the screen further away from the lenses. This virtual object is located at the original screen position, and since u and v are known, f for the diverging lens may be found.

Optical Instruments

It is assumed that the reader is already familiar with the basic structure of those optical instruments which he has studied as part of his 'O' level course.

THE CAMERA

The intensity of blackening produced on the film is a roughly proportional to the product of the exposure time and the area of the aperture. For a given contrast therefore, the time is inversely proportional to the square of the diameter of the aperture. The relative aperture is defined as the ratio $\dfrac{\text{diameter of aperture}}{\text{focal length of lens}}$ and this is denoted by an f number—usually $f/2{\cdot}8$, $f/4$, $f/5{\cdot}6$, $f/8$ etc., these numbers having been chosen because their squares are roughly in the ratio $1:2:4:8$ . . etc. It follows that opening the aperture from, say, $f/2{\cdot}8$ to $f/4$ roughly doubles the area through which light may enter the camera.

The Depth of Focus is the distance through which the film may be moved—(nearer or further from the lens) without spoiling the definition of the image.

The Depth of Field is the distance through which an object may be moved without spoiling the definition of the image. The depth of field is much greater when the aperture is small than when it is large.

MAGNIFICATION

The **magnification** of an image is given by v/u. Multiplying the lens equation throughout by v gives $v/u + v/v = v/f$ from which magnification, $m = v/f - 1$.

ANGULAR MAGNIFICATION (MAGNIFYING POWER)

A more important quantity when dealing with microscopes, telescopes etc. is the **angular magnification.** This is defined in terms of angles, because it is not always possible to know the position of the object in relation to, say, an astronomical telescope. The visual angle is the angle subtended at the eye by an object or image. Microscopes and telescopes increase the visual angle and the angular magnification is defined as:

$\dfrac{\text{Angle subtended at the eye by the image}}{\text{Angle subtended at the eye by the object}}$

THE SIMPLE MICROSCOPE

Suppose an object of height h is placed at the near point of the observer's eye. It will subtend a visual angle of $\alpha = h/D$ where D is the distance of the near point from the eye. If a converging lens is now used as a simple microscope, the observer places it so that the image is still formed at the near point, but the image now subtends a larger visual angle α^1. α^1 is given by h^1/D where h^1 is the height of the image.

Hence the angular magnification, $\dfrac{\alpha^1}{\alpha} = \dfrac{h^1}{h}$.

This is equal to the linear magnification, which, as previously explained is $v/f - 1$. Hence angular magnification is $D/f - 1$.

N.B. The image is **virtual.** If the near point is 25 cm from the eye and the focal length of the lens is 5 cm, then the angular magnification is given by $-25/5 - 1$ equals **6** numerically.

THE COMPOUND MICROSCOPE

The object is placed just outside the principal focus of the objective, a converging lens of short focal length, f_1. An intermediate image is thus formed v cm from the lens, and this image is real. The intermediate image is formed just inside the principal focus of the eyepiece lens, and the tube length (separation of the lenses) is adjusted until the final image is formed at the near point. The final image is virtual, and is inverted with respect to the object. If the focal length of the eyepiece lens is f_2 cm, then the angular magnification it produces is $(D/f_2 - 1)$. The angular magnification produced by the objective is $(v/f_1 - 1)$ so the total angular magnification of the instrument is:

$$\left(\frac{D}{f_2} - 1\right)\ \left(\frac{v}{f_1} - 1\right)$$

When drawing a ray diagram to show the path of rays through a compound microscope, remember that in addition to construction rays, several through rays must be clearly shown.

THE ASTRONOMICAL TELESCOPE

In this case the objective lens has the longer focal length f_1 cm. It produces a real intermediate image at the principal focus of the lens, since the object is at infinity.

This intermediate image is then viewed by the eyepiece lens which has a shorter focal length, f_2 cm.

If the foci of the two lenses coincide, the telescope is said to be in **normal adjustment** and the final image is formed at infinity. The angular magnification of an astronomical telescope in normal adjustment is given by f_1/f_2 .

The Astronomical Telescope with final image at near point

$\alpha = h/f_1$ and $\alpha' = h/u$, where u is the distance between the intermediate image and the eyepiece. If the final, virtual image is formed at the near

point, D cm from the eyepiece, then $\dfrac{1}{u} - \dfrac{1}{D} = \dfrac{1}{f_2}$ where f_2 is the focal

length of the eyepiece lens, and the real is positive sign convention is

used. From this equation $u = \dfrac{f_2 . D}{f_2 + D}$

Hence the angular magnification $\dfrac{\alpha'}{\alpha} = \dfrac{f_1}{u} = \dfrac{f_1(f_2 + D)}{f_2 . D}$

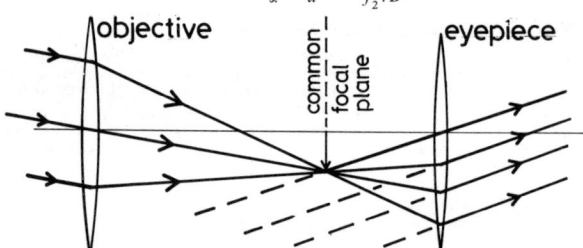

Fig. 21. Astronomical Telescope in Normal Adjustment

THE VELOCITY OF LIGHT

Most Examining Boards require the details of one method of determining the velocity of light.

Foucault's Method uses a rotating plane mirror, reflecting light from a powerful source to a concave mirror. The plane mirror is placed at the centre of curvature of the concave mirror causing the light to be reflected back along its original path. The displacement of the image, caused by the rotation of the plane mirror, is used to determine the velocity of light. The advantage of this method is that the total distance travelled by the light is of the order of metres rather than kilometres.

Defects of the Eye

The normal range of accommodation of the unaided eye is from its near point (nominally 25 cm from the eye) to its far point (nominally infinity).

SHORT SIGHT

In this case the eye is unable to see distant objects clearly since rays from such objects are brought to a focus in front of the retina. Correction is by a diverging lens. Suppose a short sighted person has a normal near point but his greatest distance of distinct vision is 250 cm. His spectacle lens must produce an image at his far point if he is to see objects at infinity. The image must be virtual, since it must be erect. Hence $\dfrac{1}{\infty} - \dfrac{1}{250} = \dfrac{1}{f}$ so that $f = -250$ cm.

LONG SIGHT

A person whose far point is normal, but who cannot clearly see near objects is said to have long sight, and correction is by a suitable converging lens. Suppose a long sighted person cannot see objects closer to his eye than 100 cm. To enable him to see clearly an object 25 cm from his eye, he must wear a spectacle which produces a virtual image 100 cm from his eye.

Hence $\dfrac{1}{25} - \dfrac{1}{100} = \dfrac{1}{f}$ so that $f = 33\frac{1}{3}$ cm (Power = +3 dioptres).

FAR SIGHT

Rays from a distant object are refracted to form a clear image behind the retina, and the far point is virtual. The rays must be additionally refracted, so that the correcting lens is again a converging one. A far sighted eye may require different lenses, one to bring the far point to infinity and one for viewing near objects. This may be achieved with the aid of a bifocal spectacle lens.

ASTIGMATISM

This is due to the eye having different powers of refraction in different planes, usually because the cornea is imperfect. It may be corrected by the use of a suitable cylindrical lens which increases the curvature in the deficient plane.

Wave Theory of Light

HUYGHENS' PRINCIPLE

Huyghens used a construction to explain the propagation of light which can also be used to explain the laws of reflection and refraction. Huyghens suggested that each point on the surface of a wave acts as a source of secondary wavelets and the wavefront X_1Y_1 is given by the envelope of the secondary wavelets.

HUYGHENS' EXPLANATION OF REFRACTION

AB is the incident plane wavefront. As each point on the wave strikes the surface it acts as a source of secondary wavelets and in the time t which it takes for light to travel the distance BC in the vacuum, it travels AD in the medium. The wavefront DC (a tangent to the wavelet at D) corresponds to the light having travelled from A to D in time t.

Hence $AD = v.t$ and $BC = c.t$.

From geometry: $BAC = i$ and $ACD = r$.

Hence $BC = AC.\sin i$ and $AD = AC.\sin r$.

Therefore $\dfrac{\sin i}{\sin r} = \dfrac{c.t}{v.t} = \dfrac{c}{v}$

Fig. 22. Huyghens' Construction for Circular Waves

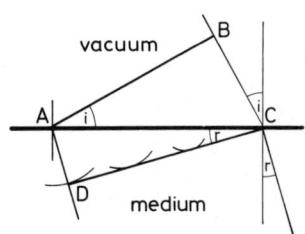

Fig. 23. Wave theory explanation of Refraction

Interference of Light

CONDITIONS FOR INTERFERENCE

Two sources of light will produce an interference pattern if:
1. They are coherent—that is they must have the same wavelength and amplitude and must be in phase with one another.
2. Their distance apart is of the order of the wavelength of light.

To all intents and purposes the two sources must derive from a single source.

YOUNG'S SLITS

Two narrow slits, S_1 and S_2, are placed in front of a single slit which is strongly illuminated by a monochromatic source. An interference pattern consisting of alternate dark and bright strips parallel to the slits is found on the screen. At O, symmetrical with respect to the two slits waves from S_1 and S_2 arrive simultaneously, reinforce each other and a bright patch is observed. At P another bright patch is observed and this occurs because the path differences S_2P-S_1P is exactly equal to one wavelength.

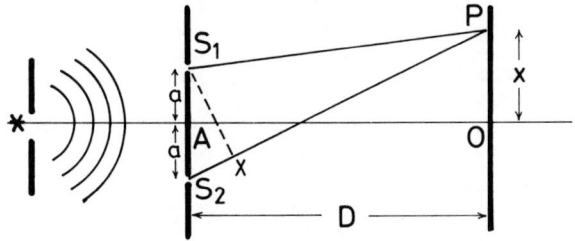

Fig. 24. Young's Slits

In order to obtain the fringes the distance D must be large compared with x. Hence the triangles APO and S_1S_2X are similar

$$\therefore \quad \frac{S_2X}{S_1S_2} = \frac{OP}{OA} = \frac{\lambda}{2a} = \frac{x}{D}$$

$$\therefore \quad \lambda = \frac{2a \cdot x}{D}$$

The second bright fringe occurs when the path difference is $2 . \lambda$, and so on. At the nth bright fringe $n\lambda = 2 \dfrac{a . x_1}{D}$ where x_1 is the distance from the centre of the pattern to the centre of the nth fringe. At the $(n + 1)$th bright fringe $(n + 1) . \lambda = 2 \dfrac{a . x_2}{D}$. Subtracting these equations

$$\lambda = 2 \frac{a}{D} (x_2 - x_1).$$

This additional theory is necessary because it is impossible to know which of the fringes is the central one. The fringes are seen to be equally spaced and to determine the wavelength of light the average separation of a number of fringes is determined.

It is not necessary to use monochromatic light to see Young's fringes—a white straight filament car headlamp bulb is adequate, but more fringes are usually seen with a monochromatic source since complex interference patterns involving colours will occur if white light is used. It is because the eye is more sensitive to yellow light that a few fringes can be seen with white light.

Polarisation

Polarisation is an important property of light waves since it helps to establish the transverse nature of the wave motion. Light consists of an electric field which vibrates at right angles to the direction of propagation, with a magnetic field mutually perpendicular with these two planes. Ordinary light is emitted by sources in which the atoms vibrate at random and the electric field occurs in all possible orientations. Such light is said to be unpolarised.

If unpolarised light passes through a sheet of Polaroid, the electric vectors are rotated so that they all lie in one plane. The light is said to be plane polarised and the plane in which the electric vectors all lie is called the plane of polarisation. The insertion of a second sheet of Polaroid in the path of the light will cause no further effect if the two sheets are parallel to each other, but as the second sheet is rotated with respect to the first the light gradually gets fainter, until, when the polarising planes of the two sheets are at right angles to each other, total extinction occurs.

NEWTON'S RINGS

An alternative way of determining the wavelength of light by an interference method is by Newton's Rings. The lens L has a very long focal length and rests on a sheet of plane glass, P. The system is illuminated by monochromatic light via the glass plate, G. A travelling microscope M is arranged to view the rings. The interference pattern is a series of concentric rings formed by interference between light reflected from, e.g. X and Y. Notice that the centre of the pattern is a **dark ring,** because the light suffers a phase change of $\lambda/2$ when it is reflected from an optically denser surface. Hence if the path difference, $2XY$, is equal to an integral number of wavelengths a dark ring is seen, and a bright ring corresponds to the condition $(n \pm \frac{1}{2}).\lambda = 2XY$.

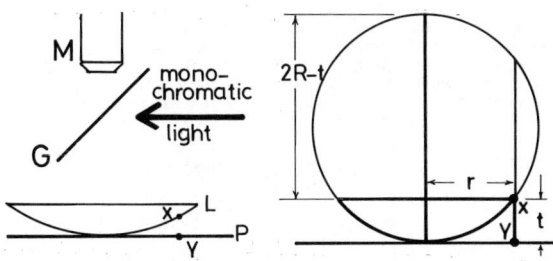

Fig. 25. Newton's Rings   Fig. 26. Theory of Newton's Rings

Determination of wavelength by Newton's Rings

Using the travelling microscope the diameter of, say, the 10th bright ring is determined. The radius of curvature of the lower lens surface must also be found.

Extending the lower lens surface to form a complete circle, and using the intersecting chord theorem, $r^2 = t(2R - t)$. But t^2 is very small by comparison with $2tR$, so to a very good approximation $2t = r^2/R$. The first bright ring must correspond to the case $n = 0$ in the equation $(n \pm \frac{1}{2}).\lambda = 2t$, so if the diameter of the 10th bright ring, r, is found, then

$$\frac{r^2}{R} = 9\frac{1}{2}\lambda.$$

As R is large, it is not accurate to use a spherometer to measure it. It should be found by obtaining rings with light of known wavelength.

Diffraction

If a plane wave is incident on a narrow slit, the slit acts as a secondary source which means that light will spread out in all directions from the slit. For a slit 100 wavelengths in aperture the effect, which causes the shadow of the slit to be larger than expected, is very small, but if the slit's width is of the order of the wavelength, then the effect is very pronounced. The same effect occurs if a very thin wire is illuminated and is known as **diffraction.**

The **diffraction grating** enables the wavelength of light to be determined very conveniently. The grating has a large number of parallel slits ruled on it, usually about 6000 per cm. $(a + b)$ is known as the grating spacing and this must be known. The grating is placed on a spectrometer so that it is perpendicular to the incident light. If the telescope is rotated a bright image of the collimator slit will be observed at various points on either side of the straight-through position. Monochromatic light should be used. The telescope is rotated to locate the first order image. This occurs when the path difference between light from adjacent slits is equal to 1 wavelength.

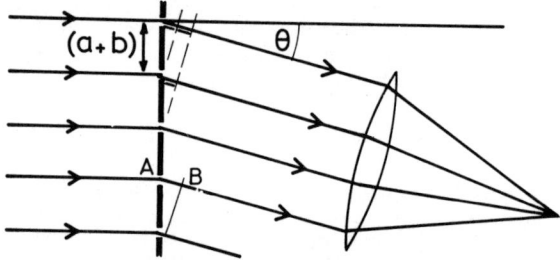

Fig. 27. The Diffraction Grating

From the above diagram it will be seen that this first order image occurs when $AB = \lambda = (a + b)\sin\theta$. There will be a second order image when $AB = 2\lambda$ and there may be a third order.

To avoid the necessity of making the grating exactly perpendicular to the incident light, it is better to find the two first order images, one on either side of the straight through position, and then halve the angle between them to find θ.

Sound Waves

Suppose a source of sound—for example the prong of a tuning fork—vibrates with Simple Harmonic Motion (see p. 14). The equation of motion of the prong is given by $y = a \sin 2\pi ft$ where y is the displacement of the prong from its mean position at time t, a is the maximum displacement (amplitude), and f is the frequency of the vibrations.

As the prong vibrates in air, it sets up a sound wave consisting of alternate compressions and rarefactions, which is propagated away from the fork. The wavelength of this sound wave is equal to the distance between successive compressions, and is denoted by λ, and the sound is propagated with a velocity v. If T is the period of the prong, then the wave is propagated through 1 wavelength in time T sec. Hence $v = \lambda/T$. But T is equal to $1/f$, so that the velocity, wavelength and frequency of the wave are related by the equation $v = \lambda . f$. Sound is an example of a **longitudinal wave motion** since the vibrations take place in directions parallel to the direction of propagation.

THE VELOCITY OF SOUND

The velocity of sound in a given medium is determined by the properties of the medium. In general $v = \sqrt{\dfrac{E}{\rho}}$ where E is the modulus of elasticity of the medium and ρ is the density.

For a solid E is equal to **Young's Modulus.**

For a gas and a liquid E is the **Bulk Modulus.**

In the case of a gas E is equal to $\gamma . p$ where p is the pressure of the gas in N.M² and γ is the ratio of the specific heats of the gas. Hence

$$v = \sqrt{\frac{\gamma . p}{\rho}}$$

For an ideal gas, which obeys Boyle's Law, $p \propto 1/v \propto \rho$ at constant temperature.

Therefore $\dfrac{p}{\rho}$ is constant and the velocity of sound in such a gas is independent of the pressure, since γ is also constant. From the general gas equation, the product of the pressure and volume of a given mass of gas is directly proportional to the absolute temperature. Since the density of the gas is inversely proportional to the volume, $p/\rho \propto T$.

Hence the velocity of sound in a gas is directly proportional to the square root of the absolute temperature.

CHARACTERISTICS OF A NOTE

1. The pitch of the note is determined by its frequency. Pitch however, is a subjective quantity whereas frequency does not depend on the observer.
2. The intensity of a note is a measure of the amplitude of the wave, the intensity being proportional to the square of the amplitude. Loudness is related to intensity, but again loudness is a subjective quantity.
3. The quality of the note is determined by the wave form. This depends on the presence of overtones, notes whose frequencies are simple multiples of the fundamental frequency.

THE DOPPLER EFFECT

This is observed when a source and observer move relative to each other. In the case of sound, a change in pitch is noted, and this is particularly noticeable when a source of sound passes an observer (e.g. when an ambulance sounding its two-tone horn passes an observer, the observer hears a sudden drop in pitch at the instant of passing).

The effect is also noted with light waves, and accounts for the red shift in astronomy.

In the examples below, f is the true frequency of the note and f' is the apparent frequency. v is the velocity of sound in air. U is the velocity of the source and V is the velocity of the observer.

The apparent frequency, f', is given by $f' = v'/\lambda'$ where v' is the apparent velocity of sound in air, and λ' is the apparent wavelength. If the source moves, then λ' varies, and if the observer moves, then v' varies.

1. If the observer is stationary and the source moves towards him, then

$\lambda' = (v - U)/f$. The apparent frequency is given by $f' = \dfrac{v \cdot f}{(v - U)}$

2. If the source is stationary and the observer moves towards it, then

$v' = v + V$. The apparent frequency is given by $f' = \dfrac{f \cdot (v + V)}{v}$

3. If both source and observer are moving towards each other, then

$v' = v + V$, and $\lambda' = (v - U)/f$. Combining these two, $f' = f \cdot \dfrac{(v + V)}{(v - U)}$.

STATIONARY WAVES

Stationary waves are produced by the combined effect of two identical waves travelling in opposite directions. A stationary wave is made up of a series of loops that are fixed in space which are called **anti-nodes.** Midway between the anti-nodes are the positions where there is no vibration and these are called **nodes.** The distance between two successive nodes (or anti-nodes) is equal to half the wavelength of the wave.

STATIONARY WAVES IN STRINGS

If a string or thin wire is fixed at its ends and plucked then transverse waves travel in each direction back and forth along the string. A stationary wave is set up with nodes at each fixed end. The string will usually vibrate in the simplest manner, with an anti-node at the centre, having the **fundamental frequency,** f.

The velocity of the wave is $V = \sqrt{\dfrac{T}{m}}$ where T is the tension in the string and m is the mass per unit length.

If l is the length of the string then $l = \lambda/2$

$$V = f\lambda = \sqrt{\frac{T}{m}} \qquad \text{hence} \qquad f = \frac{1}{\lambda}\sqrt{\frac{T}{m}} = \frac{1}{2l}\sqrt{\frac{T}{m}}$$

THE SONOMETER

The sonometer is used to investigate the vibrations of a stretched string or wire. Various forms are available, but there is a means of varying the length of the wire—usually by means of a moveable bridge—and a means of measuring the tension. This is often achieved by a screw, and a spring balance is placed between the screw and the wire, so that the tension may be determined.

Different wires may be used in order to vary m.

Various experiments can be carried out to establish the formula.

In each one only two of the four variable are altered, the other two must be kept constant.

Stationary Waves in Pipes

CLOSED PIPES

A stationary wave is set up by the interaction of the incident wave and the wave reflected from the closed end. The open end of the pipe is an anti-node and the closed end of the pipe is a node. When the air in the pipe vibrates at its fundamental frequency, the length of the pipe is $\lambda/4$. Overtones may occur in which case the length of the pipe is $3\lambda/4$, $5\lambda/4$, etc. If f_0 is the frequency of the fundamental note and v the velocity of sound in the air of the pipe, then $V = 4f_0(l + c)$ where c is the end correction.

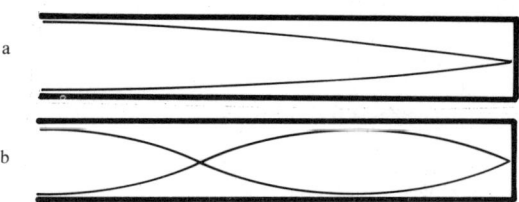

Fig. 28. Stationary Wave in Closed Pipe

OPEN PIPES

In this case there are anti-nodes at either end of the pipe. The fundamental note corresponds to a pipe length of $\lambda/2$. The overtones will correspond to pipe lengths of λ, $3\lambda/2$, etc. In this case $V = 2f_0(l + 2c)$ where c is the end correction.

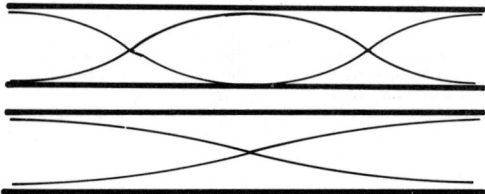

Fig. 29. Stationary Wave in Open Pipe

Resonance

Resonance is a property of all oscillating systems, but it is mentioned here as it provides the basis for a simple method of determining the velocity of sound in air.

If a tuning fork is held above a closed pipe, then the vibrations of the fork set the air in the pipe into **forced vibration.** These forced vibrations are normally of very small amplitude, but if the natural frequency of the tube is the same as that of the fork, the amplitude becomes very much larger and **resonance** is said to occur.

All oscillating systems have a natural frequency and can be made to resonate by the application of a periodic force whose frequency is equal to that of the system.

The **resonance tube** is simply a closed pipe whose length may easily be varied. This is usually achieved with water. The tube has an **end correction,** c, which must be eliminated if the velocity of sound is required. If only one fork is available it is held above the tube and the length of the air column is increased from zero until the first position of resonance is located. At this point the sound emitted by the air in tube is of maximum loudness and corresponds to Fig. 28. The length of the air column is noted, l_1. The air column is increased until a second position of resonance is found—Fig.28b. The length is again noted.

Then $l_1 + c = \dfrac{\lambda}{4}$ and $l_2 + c = \dfrac{3\lambda}{4}$

By subtraction $(l_2 - l_1) = \frac{1}{2}.\lambda$.

If f is the frequency of the fork, the velocity of sound is given by $v = f.\lambda = f.2(l_2 - l_1)$.

If a number of forks are available, the first resonance position (fundamental) for each of the forks in turn is found. A graph is plotted of l against $1/f$.

The negative intercept of the line on the $1/f$ axis is equal to the end correction and $v = 4 \times$ slope. The end correction is a complex quantity which depends on the diameter of the tube. It is approximately equal to $0.6r$, where r is the radius of the tube.

When determining the velocity of sound, the temperature of the air in the column should be noted. The velocity may then be corrected to $0°C$, using the fact that the velocity of sound is proportional to the square root of the absolute temperature.

Current Electricity

BASIC UNITS AND DEFINITIONS

The ampere is the strength of that constant current which, flowing through two parallel, straight, infinitely long conductors of negligible circular cross-section placed *in vacuo* at a distance of 1 metre apart, produces between them a force of 2×10^{-7} newton per metre of their length.

The coulomb is the quantity of electric charge transported in 1 second by a current of 1 ampere.

charge (Q) = current (I) × time (t)

The potential difference between two points of a conducting wire is 1 volt if 1 joule of work is done in transferring 1 coulomb of charge from one point to the other.

potential difference (V) = work (W) ÷ charge (Q)

The resistance of a conductor is defined by the ratio of the potential difference between its ends to the current flowing through it.

resistance (R) = potential difference (V) ÷ current (I)

Ohm's law states that the ratio V/I is a constant provided that the physical condition of the conductor does not alter. The law holds for metallic conductors, but not for non-metals. Since $W = QV$, and $Q = It$, it follows that $W = IVt$, i.e. the electrical energy, in joules, available in a circuit is equal to the product of the current in amperes, the potential difference in volts and the time in seconds. This result has already been used in the section on heat. **Power = work/time,** so the power in watts is equal to the product of the current in amperes and the potential difference in volts.

It is useful to remember that alternative expressions for energy and power are:

$$W = IVt = I^2Rt = V^2t/R$$

$$P = IV = I^2R = V^2/R$$

The resistivity of a material ρ, may be defined from the equation $R = \rho \, \dfrac{l}{A}$, where R is the resistance of a conductor of length l and cross-sectional area A. Dimensionally it will be seen that the units of resistivity are *ohm m*.

The electromotive force of a cell or generator is numerically equal to the energy given to each unit of charge as it passes through the source.

E.M.F. (E) = *energy* (W) ÷ *charge* (Q)

The e.m.f. may be measured as the potential difference between the terminals of the generator when providing no current.

Electrical generators of all kinds possess internal resistance, and some of the potential difference that they develop is used to drive current through the generator itself. If r is the internal resistance and R the resistance of the external circuit, $I = E/(R + r)$. If V is the potential difference across the external resistance (terminal P.D.), then $V = IR$. Hence:

$$I = \frac{V}{R} = \frac{E}{R + r} \qquad \therefore \ V = E \frac{R}{R + r} \qquad \text{i.e.} \quad V < E$$

CELLS IN SERIES AND PARALLEL

For a number of cells in series, $E = E_1 + E_2 + E_3 \ldots$ *etc.*, and $r = r_1 + r_2 + r_3 \ldots$ *etc.* The current through each cell is, of course, the same.

For cells of the same type in parallel, the total e.m.f. is E. The total internal resistance, X, is given by $\dfrac{1}{X} = \dfrac{1}{r} + \dfrac{1}{r} \ldots$ *etc.*

If the cells are not of the same type, no simple formula is applicable, and problems may be solved by the application of

Kirchhoff's laws.

KIRCHHOFF'S LAWS

1. **The algebraic sum of the currents to any point in a circuit is zero.**
2. **In any closed circuit the algebraic sum of the e.m.f.'s is equal to the algebraic sum of the (IR) products in each part of the circuit.**

The use of Kirchhoff's laws is best illustrated by an example.

Example. Two cells, X and Y, of e.m.f.'s 2·0 and 1·5 volts and internal resistances 1·0 and 0·5 ohm respectively are connected in parallel with a 2 ohm resistor, the cells having their positive terminals connected to the same terminal of the resistor. Find the potential difference across the resistor.

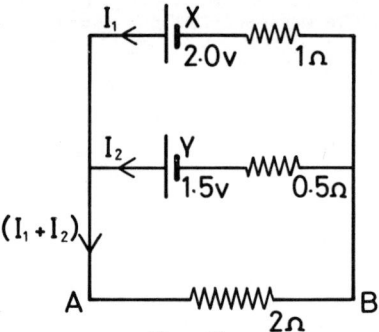

Figure 30

Answer. Applying Kirchhoff's first law to either junction, the current through the resistor is $(I_1 + I_2)$, and flows in the direction shown.

Applying Kirchhoff's second law to the circuit $XABX$:

$E = I_1 r + (I_1 + I_2)R$

$2 = 1I_1 + 2(I_1 + I_2)$ ∴ $2 = 3I_1 + 2I_2$

Applying Kirchhoff's second law to the circuit YABY:

$1 \cdot 5 = 0 \cdot 5I_2 + 2(I_1 + I_2)$ ∴ $3 = 4I_1 + 5I_2$

I_1 and I_2 may be found from this pair of simulataneous equations, and hence V may be calculated.

A MODEL FOR CONDUCTION IN METALS

The metallic bonding which occurs in metals enables their structure to be regarded as one consisting of free electrons in a matrix of positive ions. When a P.D. is applied to the ends of a metal wire, **the electrons drift** down the wire with a velocity v_d, this **drift velocity** being superimposed on their random motion. If there are n free electrons per unit volume, then the number in a wire of length l and cross-sectional area A is nAl. The total charge, Q, is $nAle$, where e is the charge on an electron. The time t taken for this charge to drift the length of the wire is l/v_d.

The current, $I = \dfrac{Q}{t} = \dfrac{nAle}{l/v_d} = nAv_de$.   Hence $v_d = \dfrac{I}{nAe}$

In many cases each atom contributes a single free electron, so a knowledge of **Avogadro's number** enables n to be found.

TEMPERATURE COEFFICIENT OF RESISTANCE

The resistance of pure metals increases with temperature, since increased thermal vibration of the ions inhibits electron flow.

$$R_\theta = R_0(1 + \alpha\theta + \beta\theta^2 + \cdots)$$

To a first approximation $R_\theta = R_0(1 + \alpha\theta)$, where α is the temperature coefficient of resistance. Hence:

$$\alpha = \frac{R_\theta - R_0}{R_0 . \theta}$$

THE POTENTIOMETER

A driving accumulator is connected across the ends of a uniform resistance wire, AB, so that there is a regular potential fall down the wire. The wire is often 1 metre long, so the potential fall is V volts per metre. The basic principle involves locating a point on the wire so that **no current flows** through the galvanometer, the point on the wire having the same potential as the plate X.

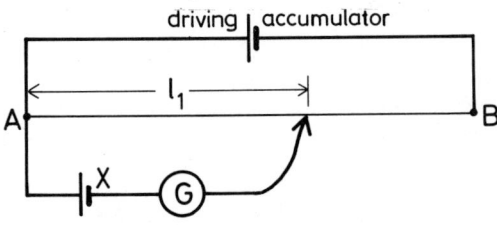

Figure 31

A balance point can only be found if:
1. Both cell and accumulator have the same plate connected to end A of the wire.
2. The e.m.f. of the cell is less than V volts.

To measure the e.m.f. of the cell, the balance length, l_1 is found. The cell is replaced with a standard cell and the new balance length, l_2, found. Then:

$$\frac{\text{e.m.f. of test cell}}{\text{e.m.f. of standard cell}} = \frac{l_1}{l_2}$$

Measurement of e.m.f. and internal resistance

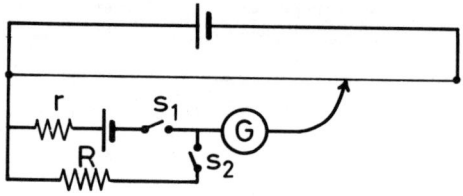

Figure 32

With S_2 open and S_1 closed, the e.m.f. E is balanced by the potential drop down the wire.

With both S_1 and S_2 closed, the cell drives current through R so that it is the terminal P.D., V, which is balanced. A shorter length of wire is needed since $V < E$.

$$V = E - Ir = E - \frac{rV}{R} \qquad \therefore \frac{1}{V} = \frac{r}{E}\cdot\frac{1}{R} + \frac{1}{E}$$

Assuming E and r to be constant, a graph of $1/V$ against $1/R$ will be a straight line with intercept $1/E$ and slope r/E. If E is found from the intercept, r may be calculated from the slope.

Use of the potentiometer to measure current

This is done by determining the value of V when the current flows through a standard resistor of resistance R. I is then calculated from $V/I = R$.

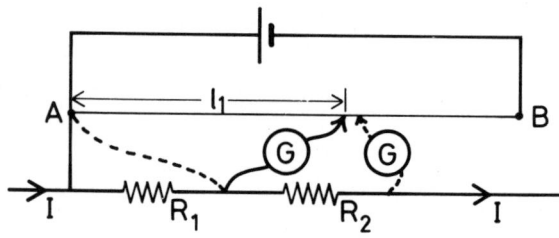

Figure 33

72

Use of the potentiometer to compare resistances

This is done using the circuit shown on the previous page. The potential drop across R_1 is first balanced against the P.D. down l_1 cm of wire. The galvanometer is moved and the connection made from A to the centre of the resistors instead of as shown. The potential drop across R_2 is thus equated to l_2 cm of wire.

Since the current through each resistor is the same:
$$\frac{V_1}{V_2} = \frac{R_1}{R_2} = \frac{l_1}{l_2}$$

As with other potentiometer measurements, for greatest accuracy l_1 and l_2 should be as large as possible, so that the percentage error in their measurement is minimised.

Measurement of small e.m.f.

Typical thermoelectric e.m.f.'s are of the order of a few millivolts. To measure these a high resistor is placed in series with the potentiometer wire. The principle of the method may be illustrated by means of a numerical example.

Suppose a driving accumulator of terminal P.D. 2 volts is used with a potentiometer wire of resistance 5Ω and length 1 m. The potential gradient is therefore 2 V m^{-1}. But if a resistor of resistance 995Ω is connected in series with the wire, the P.D. across the wire is 10 mV, and the new potential gradient is 10 mV m^{-1}. Thus e.m.f.'s up to 10 mV may conveniently be compared. Changing the value of the series resistor changes the sensitivity of the system. It may, however be necessary to increase the sensitivity of the galvanometer in order to detect the balance point with accuracy.

THE WHEATSTONE BRIDGE

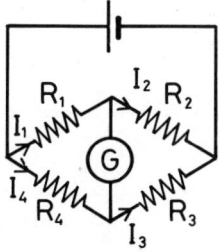

Figure 34

The potentiometer method of comparing resistances is most suitable for resistors of resistance less than 1 ohm. For higher resistances, the Wheatsone bridge is better.

If I_1, I_2, I_3 and I_4 are the currents through the four resistors, and V_1, V_2, V_3 and V_4 are the potential differences across them respectively, then for balance:

$I_1 = I_2$ and $I_3 = I_4$ since no current flows through the meter.

$\therefore V_1 = V_4$ $\therefore I_1 R_1 = I_4 R_4$

and $V_2 = V_3$ $\therefore I_2 R_2 = I_3 R_3$

$$\therefore \frac{R_1}{R_2} = \frac{R_4}{R_3} = \frac{l_1}{l_2}$$

For greatest sensitivity, the resistances in the four arms should be of the same order of magnitude. This means that with the metre bridge the balance point should be close to the centre of the wire.

AMMETERS AND VOLTMETERS

An ammeter must have a low resistance since it is connected in series in a circuit, and the P.D. across it must be kept small to limit the effect of the meter on the current which is to be measured. A voltmeter is connected in parallel with the device across which the P.D. is to be measured. **The voltmeter must have a high resistance,** otherwise the current through it will not be negligible.

The range of a meter may be extended by the use of shunts in the case of ammeters and series resistors in the case of voltmeters.

Example. A galvanometer has a resistance of 5Ω and gives a full scale deflection for a current of 15 mA. How could it be adapted to give a full scale deflection of (i) 15 A, (ii) 6 V?

(i) Only 15 mA may pass through the meter without causing damage. In a circuit carrying 15 A, 14·985 A must be shunted round the meter. The P.D. across the meter must equal the P.D. across the shunt since they are in parallel. Hence:

$0·015 \times 5 = 14·985 \times R$

$\therefore R$ can be calculated.

(ii) For full scale deflection the P.D. across the meter is $5 \times 15 = 75$ mV.
$\therefore$ the P.D. across the series resistor must be 5·925 V and its value, R, $= 5·925/0·015\Omega$.

ELECTROMAGNETISM

An electric current (moving charge) always has a magnetic field associated with it. A magnetic field is a region where magnetic forces exist, i.e. where a compass needle sets in a fixed direction and iron filings are induced to set in a pattern. Magnetic fields are conveniently pictured as consisting of a number of lines of flux, whose pattern is that of the filing field map, and whose direction is that in which the north-seeking pole of a magnet would set. **Maxwell's corkscrew rule** may also be used to find the direction of flux lines. It must be remembered that this rule, together with **Fleming's rules,** regard current as flowing from positive to negative, i.e. **they use conventional flow rather than electron flow.**

The force on a conductor in a magnetic field is proportional to the current and to the length of the conductor. The direction of the force is given by Fleming's left hand rule. $F \propto Il$, or $F = BIl$, where B is a constant involving the magnitude of the field. The equation $F = BIl$ is used to define B, **the magnetic flux density.** This is numerically equal to the force per unit length experienced by unit current which is at right angles to the magnetic field. The units of B are those of F/Il,

i.e. $\dfrac{N}{Am} = \dfrac{Nm}{Am^2} = \dfrac{J}{Am^2} = \dfrac{AVs}{Am^2} = \dfrac{Vs}{m^2} = \dfrac{Wb}{m^2} = T \, (tesla)$

If an electron of charge e moves with velocity v so that it travels a distance l in time t, then $v = l/t$, and the current I is e/t. Substituting in $F = BIl$,

$F = B \dfrac{e}{t} vt = Bev$, i.e. the force experienced by the electron at right angles to the field is Bev.

An extension of the formula $F = BIl$ is its use in calculating **the torque on a rectangular coil situated in a magnetic field.** Let the length of the side of the rectangle perpendicular to the field be a. The force on each wire is BIa, or $BIan$ if there are n turns. The forces on each side of the coil act at right angles to the field and constitute a couple. The torque is $BIanb \sin \theta$, where b is the other side of the coil and θ the angle between the normal to the coil and the field, so $b \sin \theta$ is the perpendicular distance between the forces. Hence the torque $T = BAnI \sin \theta$ where $A = ab = area$ of the coil.

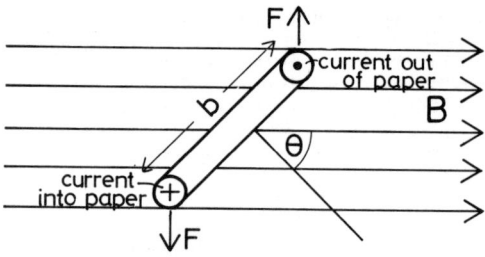

Figure 35

The coil will tend to rotate until its plane is perpendicular to the lines of magnetic flux, at which point the resultant couple is zero since the two forces are parallel to the coil. The maximum torque is $BAnI$ and occurs when the plane of the coil is parallel to the lines of magnetic flux.

$$T_{max} = BAnI$$

The product AnI is called the electromagnetic moment of the coil, which may be defined as numerically equal to the torque per unit magnetic flux density when the central axis of the coil is perpendicular to the applied magnetic field.

Since a pivoted bar magnet would also experience a torque under the same conditions, the electromagnetic moment of a bar magnet may be defined in the same way, and is equal to the ratio T_{max}/B.

In a moving coil galvanometer, the curved pole pieces and the soft iron core give a radial field, so that the plane of the coil is always parallel to the magnetic flux lines. Hence the torque is $BAnI$ and the coil rotates until this torque is equal to the restoring torque exerted by the control springs.

Hence $BAnI = c\alpha$ where c is the restoring torque per unit angular deflection and α is the angular deflection. c is normally constant so that the galvanometer has a linear scale since $I \propto \alpha$. The deflection per unit current $= \alpha/I = BnA/c$, and this is termed the sensitivity of the galvanometer.

THE BIOT-SAVART EQUATION

The Biot-Savart equation enables the magnetic flux density in different regions near to a current carrying wire to be calculated. In the general

76

case a short length of conductor, δl, carries a current of I. The product $I\delta l$ is known as a current element. The magnetic flux density at P due to this current element is given by:

$$B = \frac{\mu I \delta l \sin \theta}{4 r^2}$$

where r is the length of the line drawn from P to the centre of the short length of wire, this line making an angle θ with the wire, and μ is the permeability of the medium.

Application of the Biot-Savart equation.

For a circular coil of radius a and having n turns, the total length of wire is $2\pi an$. At all points $\sin \theta = 1$, since θ is everywhere $90°$. The total magnetic flux density at the centre:

$$B = \int \frac{\mu I \delta l \sin \theta}{4\pi r^2} = \frac{\mu I \, 2\pi \, an}{4\pi a^2} = \mu \frac{n}{2a} I$$

In the same sort of way, the magnetic flux density at a perpendicular distance a from a long straight conductor is found to be

$$B = \mu \frac{1}{2\pi a} I$$

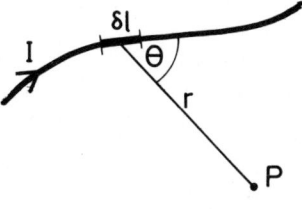

The magnetic flux density at the centre of an infinitely long solenoid of radius a and having n/l turns per metre is given by:

$$B = \mu \frac{n}{l} I$$

Figure 36

The magnetic flux density at a point on the axis of a coil is:

$$B = \mu \frac{n}{2a} I \sin^3 \alpha$$

where α is the semi angle of the cone subtended by the coil at the point.

77

ELECTROMAGNETIC INDUCTION

A conductor cutting lines of flux has an e.m.f. induced in it. The magnitude of **the induced e.m.f. is proportional to the rate at which the magnetic flux linking the conductor is changing.** A magnetic flux density B is a measure of the magnetic flux ϕ passing normally through an area A and the equation $B = \phi/A$ may thus be used to define ϕ, provided that the magnetic field is uniform. Since the unit of B, the tesla, is equivalent to *weber m^{-2}*, it follows that the unit of ϕ is the *weber (Wb)*. *Faraday's law* may be stated mathematically:

$$V = -\frac{d\phi}{dt} \quad or \quad V = -N\frac{d\phi}{dt}$$

where N is the number of conductors which cut the field. The minus sign arises from **Lenz's law** which states that **the direction of the induced e.m.f. opposes the change which causes it.** Lenz's law is a special case of the law of conservation of energy. The direction of the induced e.m.f. may be found from **Fleming's right hand rule.**

Consider a conductor being pulled with a force F along two frictionless rails so that it moves with a constant velocity v. It moves a distance x in time t such that $v = x/t$. The two rails are joined by a resistance so that a current I flows with an induced e.m.f. of V. The mechanical work done is $F.x$ and this is turned into electrical energy $I.V.t$

Thus $F.x = I.V.t$

But since the conductor is moving with constant speed the force F must be balanced by the induced force $B.I.l$, where l is the length of the conductor.

$$\therefore {}^{-}B.I.l.x = I.V.t \quad or \quad V = -B.l.x/t = -B.l.v$$

$$\therefore V = -B\frac{dA}{dt} = -\frac{d\phi}{dt}$$

A motor, when it is rotating, will also act as a generator.

The direction of the induced e.m.f. must be opposite to that of the driving potential difference (Lenz's law) otherwise the effect of the induced e.m.f. will be to accelerate the motor. The induced e.m.f. is called the back e.m.f. If the motor is held still, the current through it will be greater than when it is rotating. In the latter case the back e.m.f. limits the current; with no back e.m.f. the current is greater.

MEASUREMENT OF MAGNETIC FLUX DENSITY

A search coil of n turns and cross-sectional area A is connected to a ballistic galvanometer. The coil is placed at right angles to the magnetic field. The flux linking the coil is $n\phi = BAn$. If the flux is reduced to zero in a short time t, the induced e.m.f. in the coil is given by $E = BAn/t$. This causes an instantaneous current I to flow. $I = E/R$ where R is the total resistance of the circuit. The total charge, Q, which is circulated by the current is given by:

$$Q = It = \frac{BAn}{R}$$

A ballistic galvanometer has two essential features which make it different from a conventional moving coil galvanometer used to measure a steady current.
1. The inertia of the coil is large, and the periodic time of swing also large. This ensures that the whole of the charge passes through the coil before it starts to swing, and the first throw of the galavanometer, θ, is proportional to the total charge.
2. A ballistic galvanometer is not damped in the same way as an instrument used for measuring current. In the latter case, the coil is wound on a metal frame so that eddy currents induced in the frame quickly damp the movement of the coil. With a ballistic galvanometer the coil is wound on a wooden or plastic frame so that there is effectively no damping.

MUTUAL INDUCTANCE

If two coils are held close together, then a changing current in the primary coil induces an e.m.f. into the secondary. This e.m.f. opposes the change in the primary current, and its value is given by:

$$E = -M \frac{dI}{dt}$$

where M is the mutual inductance of the two circuits. The unit of M is the henry ($= V s A^{-1}$). **A mutual inductance of 1 henry is possessed by two circuits if an e.m.f. of 1 volt is induced in the secondary when the primary current changes uniformly at the rate of 1 A s^{-1}.**

MUTUAL INDUCTANCE OF TWO AIR-CORED SOLENOIDS

The suffixes p and s refer respectively to the primary and secondary solenoids.

$$V_s = -n_s \frac{d\phi_p}{dt} = -n_s A_p \frac{dB_p}{dt}$$

But B_p, the magnetic flux density at the centre of a long solenoid, is given by:

$$B_p = \mu_o \frac{n_p}{l_p} I_p \qquad \text{(page 77)}$$

where μ_o is the permeability of air. $\qquad \therefore V_s = -n_s A_p \mu_o \frac{n_p}{l_p} \frac{dI_p}{dt}$

But in general $V = -M \dfrac{dI}{dt}$ $\qquad \therefore M = \mu_o n_s A_p \dfrac{n_p}{l_p}$

SELF INDUCTION

If the current through a single coil changes, there is a change in the associated magnetic flux and an e.m.f. is induced in the coil itself. This induced e.m.f. opposes the change in the current and its magnitude is given by:

$$V = -L \frac{dI}{dt}$$
where L is the self-inductance of the coil.

A coil has a self inductance of one henry when the back e.m.f. in it is 1 volt, caused by the current through it charging at a steady rate of 1 A s^{-1}.

SELF INDUCTANCE OF AN AIR-CORED SOLENOID

$$V = -n \frac{d\phi}{dt} = -nA \frac{dB}{dt}$$

But as above, $B = \mu_o \dfrac{n}{l} I$, provided the solenoid is long enough for the formula to apply.

$$\therefore V = -nA\mu_o \frac{n}{l} \frac{dI}{dt} = -L \frac{dI}{dt} \qquad\qquad \therefore L = nA\mu_o \frac{n}{l} = \frac{\mu_o An^2}{l}$$

MAGNETIC PROPERTIES OF MATERIALS

An air-cored coil carrying a current has a flux density within it given by $B_o = \mu_o \cdot \mu_r H = \mu_o H$ (μ_r for air $\simeq 1$).

When a magnetic material is put inside the coil, the flux density is increased by a quantity B_i which is dependent on the type of material and on the value of H. The total Flux density within the coil is now $B = B_o + B_i$.

$$B = B_o + B_i = \mu_o H + J \qquad \text{Where } J \text{ is the Intensity of Magnetisation.}$$

$$\frac{B}{H} = \frac{B_o}{H} + \frac{J}{H} \qquad \text{or} \qquad \mu = \mu_o + \chi$$

χ is called the susceptibility of the material and is a measure of how easily the material can be magnetised. χ and μ are not constant as their values depend on H.

J–H and B–H Curves, Hysteresis

If the magnetic material is placed in a coil and the current is steadily increased, and the values of H, B and J determined, graphs of the forms shown in Figs. 37 and 38 are obtained.

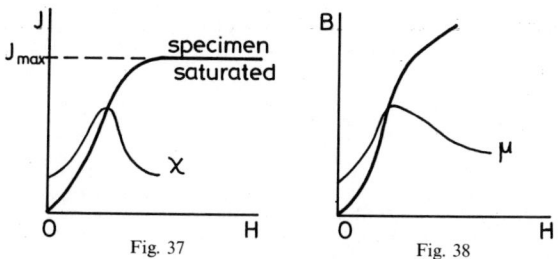

Fig. 37   Fig. 38

B does not reach saturation because $\mu_o H$ continues to increase as H increases. If the current is decreased, the magnetic path is not retraced as all the magnetic materials have the ability to retain some of their intensity of magnetisation when the magnetising force is removed.

For a complete cycle of current, i.e. increase—decrease—reverse— increase—decrease, the $B - H$ curve takes the form shown in Fig. 39 and Fig. 40.

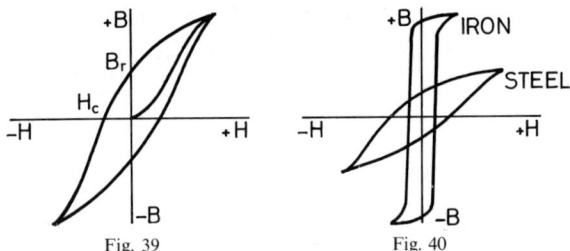

Fig. 39   Fig. 40

B_r is called the **Remanence,** i.e. **the residual flux density when the magnetising force is removed.** H_c is the **Coercivity,** or the **value of the reverse magnetising force needed** to demagnetise the material. The area of the loop represents the loss of energy in the form of heat during one magnetising cycle. **This is called the Hysteresis loss.**

If a magnetic material is to be subjected to rapid changes of magnetising force where low heating losses and a high flux density are required, e.g. in transformer cores, a material with high B_r and low H_c and low Hysteresis and eddy current losses is sought. For low eddy current losses a magnetic material alloy with high resistivity is used. Mu-metal (Ni, Fe, Cu, Cr), Permalloy and Stalloy are Examples. For permanent magnets (no cyclic operation) hysteresis losses are not important, **but high coercivity is.** Cobalt Steel (Co,W,C,Fe) and Ticonal are examples of materials used.

Magnetic properties of materials arise from the magnetic effects of moving charge (electrons) within the atoms. In a magnetic field, moving charge experiences a force and responds so as to oppose the cause (Lenz). **This is called diamagnetism.** Purely diamagnetic materials (e.g. Bi, Pb, H_2O), used as cores, reduce the magnetic effect of solenoids since, $\mu_r < 1$.

The electron orbits in some atoms are such that each atom acts as a tiny magnet and **sets in a magnetic field so as to increase the flux density.** If this effect outweighs the diamagnetism, the materials (e.g. Cu, Pt, having $\mu_r > 1$) are said to be **Paramagnetic.**

Ferromagnetic materials have crystal groupings (domains), each having a strong magnetic effect. If these are aligned, the resulting flux density is very high. Fe, Co, and Ni have $\mu_r \gg 1$.

Electrolysis

An electrolyte is a compound which, when melted or in solution, conducts electricity. Chemical changes take place at the electrodes, where reaction products may be released. In general, **hydrogen or a metal may be liberated at the cathode and oxygen or other non-metals, such as chlorine at the anode, or the material of the anode may go into solution.**

Faraday's First Law. The mass of an element liberated during electrolysis is proportional to the quantity of electricity, Q, that has flowed.

$$m = e.Q = eIt$$

where e is a constant for the element called the electrochemical equivalent. e is numerically equal to the mass deposited per unit charge (Kg/coulomb) **and is, in fact, the inverse of the ratio e/m for the ions of the element.**

Faraday's Second Law. The masses of different elements liberated by the same quantity of electricity are proportional to their chemical equivalent weights. The gm. equivalent of an element was formerly defined as the weight in gm. of the element that would combine with or displace 1 gm. of Hydrogen.

The Faraday. From Faraday's Second Law it follows that there is a quantity of electricity (F coulombs) which will liberate the gm equivalent weight of any element. This quantity is called the Faraday. $F = 96,500$ coulomb/gm equivalent.

IONIC THEORY

An atom which has more or less orbital electrons than the number required to render it electrically neutral, carries an electric charge which is a multiple of the electron charge. Such an atom is called an ion. Positive ions are short of the normal complement of electrons, whilst negative ions have extra electrons. Certain compounds are composed of ions which are held in close groupings by the electrical forces between the ions. Heating, or the introduction of a solvent can reduce the strength of the electrical bonds so that the ions become dissociated and the melt or solution has a fairly high conductivity.

A small proportion of all water molecules dissociate to OH^- and H^+ ions and these must be taken into account when considering the electrolysis of aqueous solutions.

ELECTROSTATICS

The force between two point charges Q_1 and Q_2 separated by a distance r is given by:

$$F = \frac{1}{4\pi} \frac{Q_1 Q_2}{\varepsilon r^2}$$

where ε is the **permittivity** of the medium in which the charges are situated. The permittivity of free space is 8.85×10^{-12} F m^{-1}. A charged body gives rise to an electric field in the area around it. The **electric field strength** at a point in this field, E, is numerically equal to the force acting on unit charge at the point, provided that the introduction of the charge does not affect the field. Hence $E = F/Q$, or $F = EQ$.

From the definition the units of E are seen to be N C^{-1}. E is a vector and its direction is that of the force exerted on a positive charge.

The magnitude of the electric field strength at a distance r from a charge Q situated in a medium of permittivity ε is:

$$E = \frac{1}{4\pi} \frac{Q}{\varepsilon r^2}$$

This result follows from the expression for the force between charges, since $Q_1 = Q$ and $Q_2 = 1$ coulomb by definition.

Two positive charges repel each other, and if one is moved towards the other, work has to be done. This work is stored as potential energy in the electric field. **The electric potential** at a point in a field, V, is defined as **the work done in bringing unit positive charge up from infinity to the point.**

A charge Q in a field of strength E experiences a force F. If the charge is allowed to move a small distance δx, then the work done on it by the field is $F\delta x$. In moving, its potential changes, becoming less by an amount δV as it moves from a high potential V to a lower potential $(V - \delta V)$, and the work done on it by the field is $Q(-\delta V)$. Hence:

$$F\delta x = -Q\delta V$$

$$\therefore \frac{F}{Q} = -\frac{\delta V}{\delta x} = E$$

The ratio $\delta V/\delta x$ is called **the potential gradient,** the minus sign indicating that the potential falls as x increases in the positive direction of the electric field. The potential gradient has units V m^{-1} which are also possible units for E.

Potential due to isolated charged sphere.

The potential at a point distant r from a charged sphere is equal to the work done in bringing unit charge up from infinity to the point. Thus:

$$V = \int_{x=\infty}^{x=r} \frac{(Q)(+1)(-dx)}{4\pi\varepsilon \, x^2} = \frac{1}{4\pi\varepsilon} \frac{Q}{r} \qquad Q \text{ is the charge on the sphere.}$$

The potential difference between two points at distances r_1 and r_2 from the sphere is:

$$V = \int_{x=r_2}^{x=r_1} \frac{(Q)(+1)(-dx)}{4\pi\varepsilon \, x^2} = \frac{1}{4\pi\varepsilon} Q\left(\frac{1}{r_1} - \frac{1}{r_2}\right)$$

The potential at the surface of the sphere of radius a is:

$$V = \int_{x=\infty}^{x=a} \frac{(Q)(+1)(-dx)}{4\pi\varepsilon \, x^2} = \frac{1}{4\pi\varepsilon} \frac{Q}{a}$$

CAPACITANCE

If charge is added to an isolated conductor, its potential rises but the ratio charge/potential remains constant, this constant being known as the capacitance of the conductor. If the addition of 1 coulomb of charge raises the potential of the conductor by 1 volt, then the capacitance is equal to 1 farad.

Considering a sphere of radius a, whose potential is discussed above, the capacitance of the sphere is given by:

$$C = \frac{Q}{V} = Q \div \frac{1}{4\pi\varepsilon} \frac{Q}{a} = 4\pi\varepsilon a$$

The parallel plate capacitor.

If two concentric spheres have radii r_1 and r_2 and are situated in a medium of permittivity ε, then a charge of $+Q$ on the inner sphere will induce a charge $-Q$ on the outer one. The potential difference between the spheres is:

$$V = \frac{1}{4\pi\varepsilon} Q\left(\frac{1}{r_1} - \frac{1}{r_2}\right) = \frac{1}{4\pi\varepsilon} Q\left(\frac{r_2 - r_1}{r_1 r_2}\right)$$

If r_1 and r_2 are nearly equal and very large, then $r_1 \rightleftharpoons r_2 = r$, and a section of the system may be regarded as a parallel plate capacitor.

$$C = \frac{Q}{V} = \frac{4\pi\varepsilon r^2}{d} \text{ where } d = r_2 - r_1 = \text{separation of the plates.}$$

Since the surface area of a sphere is $4\pi r^2$, the capacitance per unit area is ε/d, and the capacitance of any parallel plate capacitor is therefore

$\varepsilon A/d$, where A is the common area of the plates.

The capacitance of a parallel plate capacitor may be increased by filling the space between the plates with an insulating material. The relative permittivity of this dielectric is given by $\varepsilon_r = C/C_o$ where C is the capacitance with the dielectric between the plates and C_o is the capacitance with a vacuum between the plates. Thus:

$$\varepsilon_r = \frac{C}{C_o} = \frac{\varepsilon A/d}{\varepsilon A/d} = \frac{\varepsilon}{\varepsilon_o}$$

Capacitors in series and parallel.

For **capacitors in series,** the charge on every plate is $\pm Q$.

$V = V_1 + V_2 + V_3 \dots$ etc., so:

$$\frac{Q}{C} = \frac{Q}{C_1} + \frac{Q}{C_2} + \frac{Q}{C_3} \dots \quad \text{or} \quad \frac{1}{C} = \frac{1}{C_1} + \frac{1}{C_2} + \frac{1}{C_3} \dots$$

For **capacitors in parallel,** the potential difference across each capacitor is the same. $Q = Q_1 + Q_2 + Q_3 \dots$ etc., so:

$$CV = C_1 V + C_2 V + C_3 V \dots \quad \text{or} \quad C = C_1 + C_2 + C_3 \dots$$

Energy of a charged capacitor.

If a small additional charge δQ is placed on a capacitor having an initial charge Q at a potential V, the work done is $V\delta Q$, provided that δQ is small enough for V to be effectively constant. The total work done in placing a charge Q on a capacitor is therefore:

$$\int_0^Q V dQ = \int_0^Q \frac{Q}{C} dQ = \frac{1}{2}\frac{Q^2}{C} \quad \text{or} \quad \tfrac{1}{2}CV^2 \quad \text{or} \quad \tfrac{1}{2}QV$$

EXPERIMENTAL MEASUREMENTS

As previously mentioned, the deflection of a ballistic galvanometer is proportional to the charge passed through it, so if a capacitor is charged to a known potential and then discharged through a calibrated ballistic

galvanometer; $$C = \frac{Q}{V} = \frac{c\theta}{V}$$

For two capacitors charged to the same potential:

$$\frac{C_1}{C_2} = \frac{c\theta_1}{c\theta_2} = \frac{\theta_1}{\theta_2}$$

The relative permittivity can be found in the same way, by comparing the deflection with and without the dielectric.

Electrons

DETERMINATION OF CHARGE (MILLIKAN)

Small oil drops are produced and charged by means of a spray. They fall into the gap between the plates of a parallel plate capacitor, where they are illuminated from the side so that light is reflected from the drops into a low power microscope. With no applied electric field, a drop falls under gravity and reaches a terminal velocity v_1 when:

$$upthrust + viscous\ force = weight$$
$$\tfrac{4}{3}\pi a^3 \sigma g + 6\pi \eta a v_1 = \tfrac{4}{3}\pi a^3 \rho g$$
$$\therefore\ 6\pi \eta a v_1 = \tfrac{4}{3}\pi a^3 (\rho - \sigma) g$$

$$\therefore\ a = \left[\frac{9\eta v_1}{2(\rho - \sigma)g} \right]^{\tfrac{1}{2}}$$

where a is the radius of the drop, ρ and σ the densities of oil and air respectively, and η the viscosity of air. A potential difference V is now applied across the plates, which are separated by a distance d, so as to create an electric field of intensity $E(= V/d)$ which opposes gravity. The new terminal velocity is v_2 and:

$$upthrust + viscous\ force + electric\ force = weight$$
$$\tfrac{4}{3}\pi a^3 \sigma g + 6\pi \eta a v_2 + Ee' = \tfrac{4}{3}\pi a^3 \rho g$$

where e' is the charge on the drop. Hence:

$$Ee' = \tfrac{4}{3}\pi a^3 (\rho - \sigma) g - 6\pi \eta a v_2 = 6\pi \eta\, a(v_1 - v_2)$$

$$\therefore\ e' = \frac{6\pi \eta}{E} \left[\frac{9\eta v_1}{2(\rho - \sigma)g} \right]^{1/2} (v_1 - v_2)$$

The terminal velocities may be measured by timing the fall of the drop using a scale in the eyepiece to determine the distance travelled in a measured time. The charge on the drops may be varied by using an X-ray tube or holding a radioactive source near the gap in the top plate. If e is the charge on an electron, then $e' = ne$, where n is the number of electron charges on the drop. Millikan found that the lowest value of ne was $1 \cdot 6 \times 10^{-19}$ coulomb, and all values of ne obtained were, within limits of error, simple multiples of this value.

ALTERNATING CURRENTS

If a coil of n turns each of cross-sectional area A is rotated about an axis perpendicular to a uniform magnetic field of flux density B, it will experience an induced e.m.f. When the plane of the coil is perpendicular to the field, the instantaneous induced e.m.f. is zero, because there is zero rate of change of flux linking the coil. If the coil is rotated with uniform angular velocity ω, then at a time t after passing this zero position it will have moved through an angle ωt. The component of the flux perpendicular to the coil is $AB \cos \omega t$, so the instantaneous flux linking the coil is $nAB \cos \omega t$.

$$\therefore E = -n \frac{d\phi}{dt} = -n \frac{d(AB \cos t)}{dt} = -nAB \frac{d(\cos \omega t)}{dt}$$

$$= -BAn\omega(-\sin \omega t) = BAn\omega \sin \omega t$$

The maximum induced e.m.f., E_{max} occurs when $t = 90°$, and $E_{max} = BAn\omega$.

At any other time the induced e.m.f. is given by:

$E = E_{max} \sin \omega t$

The induced current is similarly given by:

$I = I_{max} \sin \omega t$

The mean value of E or I over a complete cycle is zero, but the current does have heating, chemical and magnetic effects. The heating effect is proportional to I^2.

$I^2 = I_{max}^2 \sin^2 \omega t = I_{max}^2 \cdot \frac{1}{2}(1 - \cos 2\omega t)$

The mean value of $\cos 2\omega t$ for 1 complete cycle is zero, so that the mean value of I^2 is $\frac{1}{2}I_{max}^2$.

$\sqrt{\dfrac{I_{max}^2}{2}} = \dfrac{I_{max}}{\sqrt{2}}$ is called the **root mean square current**, $I_{r.m.s}$

Alternating current and pure resistance.

The current and e.m.f are in phase, i.e. they are both at a maximum or a minimum at the same time.

$$\frac{V}{I} = \frac{V_{max} \sin \omega t}{I_{max} \sin \omega t} = \frac{V_{max}}{I_{max}} = R$$

Alternating current and pure inductance.

If a current $I = I_{max} \sin \omega t$ passes through a coil of inductance L and negligible resistance, the changing current sets up a back e.m.f. such that $V' = -L \, dI/dt$.

$$\therefore \ V' = -L \frac{d(I_{max} \sin \omega t)}{dt} = -L\omega I_{max} \cos \omega t$$

To maintain the current, the supply voltage must be equal in magnitude and opposite in direction to the back e.m.f.:

$$V = -V' = L\omega \, I_{max} \cos \omega t$$

The factor $L\omega$ is called **the inductive reactance** of the coil. It is measured in ohms and increases with increasing frequency. The above equation also suggests that there will be a phase difference between the current and the applied voltage. The back e.m.f. is at a maximum when the current is zero, since this point corresponds to the steepest part of the current/time graph where the current is changing most rapidly. **The current through the inductor lags $\pi/2$ behind the applied voltage.**

Alternating current and pure capacitance.

The charge Q on the plates of a capacitor of capacitance C is given by $Q = CV$. If a voltage $V = V_{max} \sin \omega t$ is supplied across a capacitor, then:

$$I = \frac{dQ}{dt} = \frac{d(CV)}{dt} = C \frac{d(V_{max} \sin \omega t)}{dt} = C\omega V_{max} \cos \omega t$$

The factor $I/C\omega$ is called the **reactance of the capacitance.** It is measured in ohms and decreases with rising frequency. Again there is a phase difference between the applied voltage and the current. The charge on the plates is at a maximum when the applied P.D. is at a maximum, and at this point the capacitor is neither charging nor discharging so that the current is a minimum. The current reaches a maximum when the capacitor is completely discharged, i.e. when the P.D. across the plates is zero. **In a capacitor, the current leads the applied voltage by a factor of $\pi/2$.**

SERIES RESONANCE CIRCUIT

In a series circuit containing resistance, capacitance and inductance, the total opposition to the flow of current comes from the resistance of the resistor and the reactances of the capacitor and inductor. Because of phase differences, these cannot simply be added together to find the total impedance of the circuit, and **a vector diagram** must be used. If V_R is taken as the base line, with the current in phase with V_R, then V_L leads by 90° on V_R and V_C lags by 90° on V_R. If $V_L > V_C$, their resultant is $(V_L - V_C)$ in the direction of V_L. By Pythagoras' theorem the applied voltage V is given by: $\quad V^2 = (V_L - V_C)^2 + V_R{}^2$

But $V_L = IX_L$, $V_C = IX_C$, and $V_R = IR$, so:

$$V^2 = (IX_L - IX_C)^2 + I^2R^2 = I^2[(X_L - X_C)^2 + R^2]$$

$$\therefore I = \frac{V}{\sqrt{(X_L - X_C)^2 + R^2}}$$

In this case the current lags on the applied voltage by an angle θ such that:

$$\tan \theta = \frac{V_L - V_C}{V_R} = \frac{IX_L - IX_C}{IR} = \frac{X_L - X_C}{R}$$

The impedance, Z, is given by $Z = \sqrt{(X_L - X_C)^2 + R^2}$.

But $X_L = \omega L$ and $X_C = 1/\omega C$, and $\omega = 2\pi f$, where f is the frequency at which the current alternates. If $X_L = X_C$ the impedance of the circuit will be a minimum, and the current a maximum, For this to occur:

$$2\pi f_o L = \frac{1}{2\pi f_o C} \quad \text{i.e. } f_o = \frac{1}{2\pi \sqrt{LC}}$$

f_o is known as the **resonant frequency** of the circuit.

POWER IN A.C. CIRCUITS

Power $P = IV = I^2R$. The mean value of I^2 is $\frac{1}{2}I_{max}^2$, so for a pure resistance the mean power absorbed is $\frac{1}{2}I_{max}^2 R$.

For pure inductance the voltage leads the current by $90°$, so:

$$VI = V_{max} I_{max} \sin \omega t \cos \omega t = V_{max} I_{max} \cdot \tfrac{1}{2} \sin 2\omega t$$

The mean value of $\sin 2\omega t$ for 1 complete cycle is zero, so that the mean energy dissipated in a pure inductance is zero. In the first quarter cycle the energy is stored in the magnetic field, and in the second quarter cycle the energy is returned to the generator.

By the same argument, the mean energy dissipated in a pure capacitance is also zero, the energy being stored in the electrostatic field during the first quarter cycle, and returned to the generator during the second quarter cycle.

For the general A.C. circuit, the power dissipated is that due to resistive components, and equals I^2R. However, in the general case, I and V are not in phase, where V leads I by an angle θ, $V \cos \theta$ is the resistive component of V, and $V \sin \theta$ is the inductive component. The mean inductive power is zero, so the mean power is $IV \cos \theta$, where $\cos \theta$ is the **power factor** of the circuit.

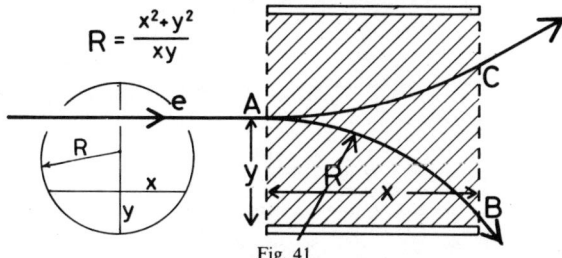

$$R = \frac{x^2 + y^2}{xy}$$

Fig. 41

The work done on an electron of charge e, accelerated through a potential V is Ve. This is approx. equal to the final K.E. of the electrons $= \frac{1}{2}mv^2$ (the thermal velocities of the electrons on leaving the cathode $\ll v$). $Ve = \frac{1}{2}mv^2$ . . . . . ①

If the electrons pass into a magnetic field perpendicular to their direction of motion, they are acted on by a force Bev where B is the flux density of the field. The direction of the force is given by the left hand rule (remembering that the electron carries negative charge) and is always perpendicular to the electrons direction of motion. The electron path is along the arc of a circle. $Bev = mv^2/R$ . . . . . ② where R is the radius of curvature. Eliminating v from the two equations:

$e/m = \dfrac{V}{R^2 B^2}$. This method is limited by the inaccuracy of equation 1.

The electrons, having passed through the anode, experience a retarding force due to the attraction on them of the anode. By employing the balanced deflections method, this equation is not used.

Balanced Deflection Method

If electrons pass through an electrostatic field perpendicular to their path, the force on them is $F = Ee$ and they follow a parabolic path. If a magnetic field is simultaneously applied so that the force due to it is in the opposite direction a balance between the forces can be obtained and the electron stream is undeflected.

$Bev = Ee$ and so $v = E/B$ *for* Balanced forces.

$Bev = \dfrac{mv^2}{R}$ *for* Magnetic Field Alone. $e/m = v/BR = E/B^2R$

Radioactivity

The nucleii of some atoms spontaneously disintegrate and radiation of three types may be emitted. These are called α, β and γ radiation and they are all able to penetrate matter and interact with it.

α **Radiation** is deflected by a magnetic field in a direction showing that it is positively charged. It causes ionisation and so can be detected in a cloud chamber, where collision effects indicate that it is made up **of particles having the same mass as a helium nucleus.** ($90°$ separation after an oblique collision), e/m experiments indicate that α particles carry a double positive electron charge. The energy of α particles from any one disintegrating material, has a discrete value and their speed is of the order of $\frac{1}{20}C$, where C is the velocity of light α **particles have a short range, the most energetic stopped by about 0.1 mm of aluminium.**

Rutherford used α particles to bombard thin metal foils and found that some were deflected by very large angles. This led him to postulate the existence of a small, heavy, positively charged, central nucleus to the atom.

β **Radiation** can also be magnetically deflected, consistent with it being negatively charged **and is identified as a stream of electrons.** β particles are emitted with a wide range of speeds (up to $0.98C$). Their range is greater than for α's, being **able to pass through up to 1 cm of aluminium.**

γ **Radiation is not deflected by a magnetic field.** It has the same general properties as light, modified in degree by the very much shorter wavelengths. They have very much greater penetrating ability than either α or β radiations.

Radioactive Decay. This is a **random process** and the rate at which atoms disintegrate is proportional to the number of atoms present. $dN/dt = -\lambda N$, where λ is called the decay constant. Integration gives $N = N_o\, e^{-\lambda t}$, where N_o is the number of atoms present at time $t = 0$. The number of atoms remaining not disintegrated, decreases exponentially with time. **The Half-Life for radioactive material is the time in which half the original atoms have disintegrated.**

The Half-Life $T = \dfrac{0.693}{\lambda}$ and the mean lifetime of a single radioactive

atom $= \dfrac{1}{\lambda}$.

ABSORPTION OF α, β AND γ RAYS

When an α particle passes close to an atom, there is an electrostatic force between it and the atom's orbiting electrons, and a "collision" can occur in which one electron may be given sufficient energy to escape from the atom. An ion pair is thus created—the electron and the ionised atom. After a number of such collisions, the α particle will not have sufficient energy to cause ionisation. The distance then travelled is called the range. At this stage, the α particle captures two electrons and forms a neutral helium atom.

β particles also cause ionisation but, because they have the same mass and charge as the orbiting electrons, the "collisions" cause scattering (deflection) of the β particles. The density of ion-pairs along the path of a β particle is much less than for an α particle. They therefore travel considerably further than α particles before they have insufficient energy to cause further ionisation.

γ radiation photons (see photoelectric effect) may interact with an orbiting electron, or with a nucleus or may spontaneously produce an electron-positron pair. **Whichever absorption process occurs, it is the secondary electrons produced which cause ionisation of the medium.** The ionisation produced by the radiations can be used to detect them.

CLOUD CHAMBERS

A saturated, or super-saturated vapour will only condense into droplets, if there are suitable condensation centres present. **Dust or ionised particles provide the necessary centres.**

In the expansion cloud chamber, the sudden cooling of a vapour by expansion super-saturates it, and droplets will form along the ionisation path produced by an ionising radiation.

In the diffusion cloud chamber, super-saturation is brought about as the vapour drifts downwards towards a surface kept at a very low temperature.

Ionisation Chambers and Geiger-Muller Tubes

In these detectors, the ion pairs are produced in a region between two electrodes maintained at a high potential difference. Under the influence of the electric field, the ions accelerate and cause further ionisation by collision. In this way, an avalanche builds up and a small current pulse is created.

X-Rays

X-rays are produced by the bombardment of a heavy metal target by high energy electrons. In the Coolidge type tube the electrons are produced by a heated filament and are accelerated in high intensity electric field and strike the target which is part of the anode. The target is set at an angle to the incident electron beam and the X-rays produced pass through a side window of the tube.
Only a small proportion of the energy of the electron beam is converted to X-radiation, the rest being dissipated as heat in the anode. The anode is cooled by circulating oil through ducts in the anode.

X-rays are able to penetrate materials, though they are absorbed by about 1 mm of lead. They travel in straight lines, affect photographic emulsions, cause ionisation and photoelectric emission. They are not deflected by magnetic or electrostatic fields and so **cannot be a stream of charged particles.** Early attempts to show that the radiation was an electromagnetic wave, by showing diffraction effects, failed since the gratings used did not have sufficiently small grating spaces. Von Laue suggested that a crystal, with its regular array of atoms, very closely spaced, would act as a diffraction grating for X-rays **if the atomic spacing was of the same order as the X-ray wavelength.**

This was tried and a diffraction pattern was obtained on a photographic plate.

The spectrum of X-rays from any target consists of (a), a continuous spectrum and (b), a line spectrum. The continuous spectrum (Bremsstrahlung—braking radiation) arises from those electrons which, when approaching a nucleus, accelerate and give out radiation (losing energy in the process). The line spectrum occurs because some incident electrons are able to penetrate far enough into the atom to "collide" with electrons in the inner shells. If such electrons are removed from their shell a later transition of an electron from a higher energy, outer shell, to fill the gap, results in the release of radiation energy with a discrete wavelength.

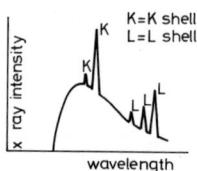

Fig. 42

The Photoelectric Effect

A negatively charged electroscope connected to a clean zinc plate is rapidly discharged when the plate is irradiated with ultra-violet light. No such effect occurs if the electroscope is positively charged. The radiation releases electrons from the zinc atoms which are held back when the plate is positively charged, but are emitted when it is negatively charged. e/m measurements prove that the emitted particles are electrons.

Similar emission from other metals can be caused by X-rays, ultra-violet, visible light and infra-red.

LAWS OF PHOTOELECTRIC EMISSION

1. The number of photoelectrons emitted/sec increases as the intensity of the incident radiation increases.

2. The photoelectron energy ranges from zero up to a maximum which increases with the radiation frequency.

3. For each metal, there is a minimum (threshold) radiation frequency below which no emission takes place however high the radiation intensity

Two of these results **are not consistent with a wave theory** of radiation. To release an electron from a metal, a certain **minimum energy is required**. Continuous waves would continuously bring energy to the metal surface and eventually electrons would be bound to be released, **whatever the frequency of the radiation.** In addition, once an electron had been released, there should be no limit to the extra energy it could collect from the incident waves and carry away as Kinetic Energy.

The observed behavior of Black-Body Radiation suggested to Planck, **that radiation was emitted in integral multiples of a quantum of energy of value** $E = hf$, where h is a constant and f is the frequency of the radiation.

Einstein showed that this quantum theory could be applied to explain the laws of photoelectric emission. A quantum of radiation energy is called a photon. Infra-red has a comparatively long (high) wavelength and hence, a low frequency and the photon energy is thus small compared with that of ultra-violet or X-rays.

Key Facts Educational Aids

KEY FACTS CARDS

Latin
Julius Caesar
New Testament
German
Macbeth
Geography Regional
English Comprehension
English Language
Economics
Elementary Mathematics
Algebra
Modern Mathematics

English History (1815–1914)
English History (1914–1946)
Chemistry
Physics
Biology
Geometry
Geography
French
Arithmetic and Trigonometry
General Science
Additional Mathematics
Technical Drawing

KEY FACTS COURSE COMPANIONS

Economics
Modern Mathematics
Algebra
Geometry
Arithmetic and Trigonometry
Additional Mathematics

Geography
French
Physics
Chemistry
English
Biology

KEY FACTS A-LEVEL BOOKS

Chemistry
Biology

Pure Mathematics
Physics

KEY FACTS O-LEVEL PASSBOOKS

Modern Mathematics
Geography
Biology
Chemistry

Physics
English History (1815–1939)
French
English

KEY FACTS O-LEVEL MODEL ANSWERS

Modern Mathematics
Geography
Biology
Chemistry

Physics
English History (1815–1939)
French
English

KEY FACTS REFERENCE LIBRARY

O-Level Biology
O-Level Physics
O-Level Chemistry

O-Level Trad. & Mod. Maths
O-Level Geography
O-Level History

KEY FACTS A-LEVEL PASSBOOKS

Physics
Biology

Chemistry
Pure Maths